Friction Stir Welding and Processing in Alloy Manufacturing

Friction Stir Welding and Processing in Alloy Manufacturing

Special Issue Editor

Carlos Agelet de Saracibar

MDPI • Basel • Beijing • Wuhan • Barcelona • Belgrade

MDPI

Special Issue Editor
Carlos Agelet de Saracibar
UPC BarcelonaTech
Spain

Editorial Office
MDPI
St. Alban-Anlage 66
4052 Basel, Switzerland

This is a reprint of articles from the Special Issue published online in the open access journal *Metals* (ISSN 2075-4701) from 2018 to 2019 (available at: https://www.mdpi.com/journal/metals/special_issues/stir_processing)

For citation purposes, cite each article independently as indicated on the article page online and as indicated below:

LastName, A.A.; LastName, B.B.; LastName, C.C. Article Title. *Journal Name* **Year**, *Article Number*, Page Range.

ISBN 978-3-03921-207-1 (Pbk)
ISBN 978-3-03921-208-8 (PDF)

Contents

About the Special Issue Editor

Carlos Agelet de Saracibar is Civil Engineer (UPC BarcelonaTech, 1985) and Civil Engineer, Ph.D. (UPC BarcelonaTech, 1990). He has been assistant professor (1988–1991), associate professor (1991–2003), and full professor of continuum mechanics and structural analysis (since 2003) at the School of Civil Engineering, UPC BarcelonaTech. He was a visiting scholar at the Division of Applied Mechanics at Stanford University (1993), invited by Prof. J.C. Simo. He has been the Vice-Dean of Research, Chair of the Research & Doctoral Committee, and Head of the Doctoral Program in Civil Engineering at the School of Civil Engineering, UPC (1997–2005), and member of the Doctoral Committee of the UPC (1998–2002). He has also been an elected member of the European Scientific Association for Material Forming (ESAFORM) Scientific Committee (1998–1999) and the ESAFORM Board of Directors (2000–2012). He is responsible for the Specific Research Center Computational Continuum Mechanics (MC)2-UPC (since 2018). He has been acting regularly as an evaluation expert for national and international governmental agencies, universities, and scientific associations and institutions; a panelist for technological institutions; and a reviewer for several international scientific journals.

Prof. Carlos Agelet de Saracibar has published 38 research papers in JCR-indexed scientific journals, 127 communications in international conferences, 10 research monographs, and five book chapters in the field of non-linear computational mechanics of solids, with a focus on finite element technology, non-linear constitutive models, contact mechanics, coupled thermo-mechanical problems, and numerical simulation of material forming and production processes, such as casting, friction stir welding (FSW), and additive manufacturing (AM). He has also published three textbooks on continuum mechanics. He has been an invited guest editor of Special Issues in the *International Journal of Plasticity,* the open access journals *Metals and Mathematics, and Comptes Rendus Mécanique.* He is member of the editorial board of *Computer Methods in Materials Science* (since 2006), *International Journal of Material Forming* (since 2008), and *Revista Internacional de Métodos Numéricos para Cálculo y Diseño en Ingeniería* (since 2010).

Prof. Carlos Agelet de Saracibar has participated in over 70 national and international conferences. He has been chairman of two international conferences, a member of the scientific committee in over 40 international conferences, and a member of the international organizing committee of seven international symposia. He has also co-organized over 35 invited mini-symposia and invited sessions. He has been invited as plenary lecturer in one international conference and as keynote lecturer in 12 international symposia and conferences. Prof. Carlos Agelet de Saracibar has participated in several European and nationally funded research projects, mainly related to the application of the finite element method (FEM) to the development of tools for computer aided engineering (CAE), with a focus on the computational modeling and numerical simulation of coupled thermo-mechanical problems, with applications to the numerical simulation of industrial metal forming processes, such as casting, laser heat forming (LHF), shaped metal deposition (SMD), welding and friction stir welding (FSW), and additive manufacturing (AM) processes.

Preface to "Friction Stir Welding and Processing in Alloy Manufacturing"

The computational modeling and numerical simulation of friction stir welding (FSW) processes are extremely challenging tasks due to the highly nonlinear and coupled nature of the physical problem and the numerical issues that need to be properly addressed. This is why the numerical simulation of FSW processes has been a very active research field over the last few decades. Despite the complexity of the physical problem and its numerical simulation, significant advances in the field have been achieved as a result of interdisciplinary research on related fields of computational mechanics, constitutive modeling, materials characterization, mathematical analysis, and numerical methods. On the other hand, also during this period, the industry has shown a growing interest in the field, incorporating predictive numerical techniques as a valuable tool for design and process optimization of FSW processes.

This MDPI book is the printed copy edition of the Special Issue (SI) "Friction Stir Welding and Processing in Alloy Manufacturing" that was previously published online in the open access journal Metals. The book collects 10 papers with the latest developments in the fields of FSW, friction stir spot welding (FSSW), and friction stir processing (FSP), written by well-known researchers who have contributed significantly to advances in computational modeling, numerical simulation, and material characterization in the field. Each contribution has been subjected to peer review by three experts in the field in order to monitor the research quality of the outcome. Sixteen contributions were submitted, with six of them being rejected.

The research topics addressed in the book include, among others, the effect and influence of different FSW process parameters, such as the effect of the tool tilt angle on the heat generation and the material flow of FSW, the influence of the pin shape on the high rotation speed of a FSW joint of an AA-6061-T6 sheet, the performance of plunge depth control methods during FSW, the effect of tool rotational speeds on the microstructure and mechanical properties of dissimilar FSW CuCrZr/CuNiCrSi butt joints, and the influence of alloy position, rolling, and welding directions on properties of AA2024/AA7050's dissimilar butt weld obtained by FSW. Other research topics on FSW addressed in the book include a correlation between the ultimate shear stress and the thickness affected by intermetallic compounds in FSW of dissimilar aluminum alloy–stainless steel joints and abnormal grain growth in the heat-affected zone of a friction stir welded joint of 32Mn-7Cr-1Mo-0.3N steel during post-weld heat treatment.

Furthermore, the book includes a study on FSSW processes on the compensation of vertical position error using a force–deflection model in friction stir spot welding, two studies on FSP, a study on another approach to characterizing particle distribution during surface composite fabrication using friction stir processing, and a study on the characterization of microstructural refinement and the hardness profile resulting from friction stir processing of 6061-T6 aluminum alloy extrusions.

I would like to acknowledge all the contributors for submitting their latest developments in the field to this SI, now a printed book. I would like also to acknowledge the great support I have always received from the MDPI team with respect to all the editorial tasks.

Carlos Agelet de Saracibar
Special Issue Editor

metals

MDPI

Editorial

Challenges to Be Tackled in the Computational Modeling and Numerical Simulation of FSW Processes

Carlos Agelet de Saracibar

Department of Civil and Environmental Engineering, School of Civil Engineering, Technical University of Catalonia, UPC BarcelonaTech, 08034 Barcelona, Spain; carlos.ageletdesaracibar@gmail.com

Received: 10 May 2019; Accepted: 15 May 2019; Published: 17 May 2019

The computational modeling and numerical simulation of Friction Stir Welding (FSW) processes is an extremely challenging task due to the highly nonlinear and coupled nature of the physical problem and the complex computational issues that need to be properly tackled in the numerical model [1–6].

1. Physical Model

The FSW process is a complex problem due to the highly nonlinear and coupled nature of the physical problem. Different physical phenomena occur during the welding process, involving the thermal and mechanical interactions. The temperature field is a function of many welding parameters such as welding speed, welding sequence and environmental conditions. Formation of distortions and residual stresses in workpieces depend on many interrelated factors such as thermal field, material properties, structural boundary conditions and welding conditions. The challenging issues in physical modeling of the FSW process are divided into four parts.

1.1. Complex Thermal Behavior

Heat transfer mechanisms including convection, radiation and conduction have a significant role on the process behavior. Convection and radiation fluxes dissipate heat significantly through the workpieces to the surrounding environment, while conduction heat flux occurs between the workpieces and the support.

1.2. Non-Linear Behavior and Localized Nature

The mechanical behavior during FSW is non-linear due to the high strain rates and visco-plastic material. The strong non-linear region is limited to a small area and the remaining part of the model is mostly linear. However, the exact boundaries of the non-linear zone are not known a priori. Knowledge of strain rate is important for understanding the subsequent evolution of grain structure, and it serves as a basis for verification of various models as well.

1.3. Coupled Nature

The thermal and mechanical problems are strongly coupled. The mechanical effects coupled to the thermal ones include internal heat generation due to plastic deformations or viscous effects, heat transfer between contacting bodies, heat generation due to friction, etc. The thermal effects are also coupled to the mechanical ones; for instance, thermal expansion, temperature-dependent mechanical properties, temperature gradients in workpieces, etc. An adequate physical model of the welding process must account for all these phenomena including thermal, mechanical and coupling aspects.

1.4. Thermo-Mechanical Frictional Contact Nature

Thermo-mechanical frictional contact between the tool and the workpieces plays a crucial role. Interactions between the contacting bodies include impenetrability, frictional stresses, heat generation due to friction and thermal conduction at the contact interface. An adequate physical model of the FSW process must properly account for all those phenomena.

2. Numerical Model

The numerical simulation of the FSW process by the Finite Element Method (FEM) has many complex and challenging aspects that are difficult to deal with. The welding process is described by the momentum and energy balance equations governing the coupled thermo-mechanical problem. Both governing equations are non-linear and this has important implications upon the complexity of the numerical model. Consequently, a robust and efficient numerical strategy is crucial for solving such highly non-linear coupled FE equations. Numerical simulation of the FSW process can be carried out at a local or global level [1,2]. In local level analysis, the focus of the simulation is the Heat Affected Zone (HAZ). The simulation is intended to compute the heat power generated either by visco-plastic dissipation or by friction at the contact interface. At this level, the relevant process phenomena are the relationship between welding parameters, the contact mechanisms in terms of applied normal pressure and friction coefficient, the setting geometry, the material flow within the HAZ, its size and the corresponding consequences on the microstructure evolution, etc. A simulation carried out at global level studies the entire component to be welded. In this case, a moving heat power source is applied to a control volume representing the actual HAZ at each time-step of the analysis. The effects induced by the welding process on the structural behavior, such as distortions, residual stresses or weaknesses along the welding line, are the target of this kind of study. The challenging issues in numerical modeling of the FSW process are divided into the following seven parts.

2.1. Mechanical Problem

The mechanical problem is governed by the momentum balance equation. A quasi-static mechanical analysis can be assumed as the inertia effects in welding processes are negligible due to the high viscosity characterization. At local level, the volumetric changes are found to be negligible, and incompressibility can be assumed. To deal with the incompressible behavior, a very convenient and common choice is to describe the formulation splitting the stress tensor into its deviatoric and volumetric parts. Dealing with the incompressible limit requires the use of mixed velocity-pressure interpolations. The problem suffers from instability if the standard Galerkin FE formulation is used, unless compatible spaces for the pressure and the velocity fields are selected (LBB stability condition). Due to this, pressure instabilities appear if equal velocity-pressure interpolations are used. Thus, the challenging issue of pressure stabilization rises up [1–3,6]. The welding process is characterized by very high strain rates as well as a wide temperature range going from the environmental temperature to the melting point. Hence, the constitutive laws adopted should depend on both variables. At typical welding temperatures, the large strain deformation is mainly visco-plastic. Depending on the scope of the analysis, rigid-visco-plastic or elasto-visco-plastic constitutive models can be used. Not only the prediction of the temperature evolution, but the accurate residual stress evaluation field generated during the process is the objective of the FSW simulation. The selected constitutive model must appropriately define the material behavior and has to be calibrated by the temperature evolution. The challenge arises from the extremely non-linear behavior of these constitutive models and, therefore, from the numerical point of view, a special treatment is obligatory. Moreover, the localized large strain rates usually involved in FSW processes make the problem even more complex.

2.2. Thermal Problem

The thermal problem is defined by the energy balance equation. In FSW simulation, the plastic dissipation term appearing in the energy equation has a critical role on the process behavior and it is the main source of internal heat generation. The definition of the heat source is one of the key points when studying the welding process. In global level simulations, the mesh density used to discretize the geometry is not usually fine enough to define the welding pool shape or a non-uniform heat source. This is only done if the simulation of the welding pool is the objective itself (local level analysis). If the global structure is considered (global level analysis), the size of the heat source is of the same dimension than the element size generally used for a thermo-mechanical analysis. Therefore, in a global level analysis the resulting mesh density is usually too coarse to represent the actual shape of the heat source. Depending on the kinematic framework used to describe the formulation of the coupled thermo-mechanical problem, a convective term might appear in the thermal governing equations. Therefore, convection instabilities of the temperature appear for convection dominated problems [3,6]. It is well known that in diffusion dominated problems, the solution is stable. However, in convection dominated problems, the stabilizing effect of the diffusion term becomes insufficient and oscillations appear in the temperature field. The threshold between stable and unstable solutions is usually expressed in terms of the Peclet number.

2.3. Kinematic Framework

Establishing an appropriate kinematic framework for the simulation of FSW processes is a key issue. If the welding process is studied at global level, the use of a Lagrangian framework is an appropriate choice for the description of the problem. The Lagrangian reference frame allows easy tracking of free surfaces and interfaces between different materials. In a local simulation, the main focus of the simulation is the HAZ where the use of a Lagrangian framework is not always advantageous. In the HAZ, the large distortions would require continuous re-meshing. The alternative is to use Eulerian or Arbitrary Lagrangian Eulerian (ALE) methods. The Eulerian formulation facilitates the treatment of large distortions in the fluid motion. Its handicap is the difficulty to follow free surfaces and interfaces between different materials or different media. An Arbitrary Lagrangian Eulerian (ALE) formulation is particularly useful in flow problems involving large distortions in the presence of mobile and deforming boundaries. In the simulation of FSW, it is adroit to introduce an apropos kinematic framework for the description of different parts of the computational domain [4,6]. Despite the efficiency of the idea, the mesh moving strategy and the treatment of the domains' interaction are challenging.

2.4. Thermo-Mechanical Frictional Contact Problem

The computational modeling of the thermo-mechanical frictional contact between the tool and workpieces is a key issue in the numerical simulation of the FSW process [1,2,6]. The computational model must accurately deal with contact impenetrability, frictional behavior, heat generated by friction and heat transfer due to thermal contact at the contact interface. Penalty-based methods, such as the penalty method or the Uzawa's version of the augmented Lagrangian method, Lagrange multipliers or direct elimination methods, can be used to model the mechanical frictional contact interaction. Within the framework of a fluid mechanics approach, a Norton thermo-frictional contact model can be used to compute the tangential component of the traction vector at the contact interface in terms of the variation of the relative slip velocity. The heat flux generated by friction at the contact interface between the tool and the workpieces can be split into two parts, that is, a part absorbed by the tool and a part absorbed by the workpieces, where the amount of heat absorbed by the tool and the workpieces depends on the thermal diffusivity of the two materials in contact. Alternatively, as a limit case, full stick thermo-mechanical contact conditions between the tool and the workpieces can be also considered. In this case, the temperature and velocity fields are continuous through the contact interface between the tool and the workpieces.

2.5. Coupled Problem

The numerical solution of the coupled thermo-mechanical problem involves the transformation of an infinite dimensional transient system into a sequence of discrete non-linear algebraic problems [6]. This is achieved by means of the FE spatial discretization procedure, a time-marching scheme for the advancement of the primary nodal variables and a time integration algorithm to update the internal variables of the constitutive equations. Regarding the time-stepping schemes, two types of strategies can be applied to the solution of the coupled thermo-mechanical problems. The first possibility is to use a monolithic (simultaneous) time-stepping algorithm which solves both the mechanical and the thermal problems together. It advances all the primary nodal variables of the problem simultaneously. The main advantage of this method is that it enables stability and convergence of the whole coupled problem. However, in simultaneous solution procedures, the time-step as well as time-stepping algorithm has to be equal for all subproblems, which may be inefficient if different time scales are involved in the thermal and the mechanical problem. Another important disadvantage is the considerably high computational effort required to solve the monolithic algebraic system and the necessity to develop software and solution methods specifically for each coupled problem. A second possibility is a staggered algorithm (block-iterative or fractional-step), where the two sub-problems are solved sequentially. Usually, a staggered solution, arising from an operator split and a product formula algorithm (PFA), yields superior computational efficiency. Staggered solutions are based on an operator split, applied to the coupled system of non-linear ordinary differential equations, and a product formula algorithm, which, within the framework of classical fractional step methods, leads to a splitting of the original monolithic problem into two smaller and better conditioned sub-problems. This leads to the partition of the original problem into smaller and typically symmetric (physical) subproblems. After this, the use of different standard time-stepping algorithms developed for the uncoupled sub-problems is straightforward, and it is possible to take advantage of the different time scales involved. The major drawback of these methods is the possible loss of accuracy and stability. However, it is possible to obtain unconditionally stable schemes using this approach, providing that the operator split preserves the underlying dissipative structure of the original problem.

2.6. Particle Tracing

One of the main issues in the study of FSW at local level is the heat generation. The generated heat must be enough to allow for the material to flow and to obtain a deep HAZ. Insufficient heat forms voids as the material is not softened enough to flow properly. The visualization of the material flow is a very useful tool to understand its behavior during the weld. It can be used to investigate the appropriate process parameters to create a qualified joint. However, following the position of the material during the welding process is not an easy task, neither experimentally (needs metallographic tools) or numerically. This is why establishing a numerical method for the visualization of the material trajectory in order to gain insight to the HAZ and the material penetration within the thickness of the workpieces is one of the key issues of the numerical simulation. Particle tracing is a method used to simulate the motion of material points, following their positions at each time-step of the analysis [5]. In the Lagrangian framework the trajectories are given by the displacement field. When using Eulerian and ALE framework the solution does not give directly information about the material position. However, the velocity field obtained can be integrated to get an insight of the extent of material mixing during the weld. Integration of the velocity field is proposed at post-process level to follow the material motion. An appropriate time integration method for the solution of the ODE in order to track the particles is needed. Moreover, a search algorithm must be executed to find the position of the material points if Eulerian or ALE meshes are used.

2.7. Residual Stresses

Generally, FSW yields fine microstructures, absence of cracking, low residual distortion, and no loss of alloying elements. Nevertheless, as in the traditional fusion welds, a softened HAZ and a tensile residual stress field appear. Although the residual stresses and distortion are smaller in comparison with those of traditional fusion welding, they cannot be ignored, especially when welding thin plates of large size. In the local level analysis, the focus of the study is the HAZ and a visco-plastic model is used to characterize the material behavior. Elastic stresses are neglected, and thus, the calculation of residual stresses is not possible. However, at global level, the residual stresses are one of the main outcomes of the process simulation using an elasto-visco-plastic constitutive model. The use of a local-global coupling strategy has been proposed as a method to obtain the residual stress field, as this a challenging issue [2].

Conflicts of Interest: The authors declare no conflicts of interest.

References

1. Dialami, N.; Chiumenti, M.; Cervera, M.; Agelet de Saracibar, C. Challenges in thermo-mechanical analysis of Friction Stir Welding processes. *Arch. Comput. Methods Eng.* **2017**, *24*, 189–225. [CrossRef]
2. Dialami, N.; Cervera, M.; Chiumenti, M.; Agelet de Saracibar, C. Local-global strategy for the prediction of residual stresses in FSW processes. *Int. J. Adv. Manuf. Technol.* **2017**, *88*, 3099–3111. [CrossRef]
3. Agelet de Saracibar, C.; Chiumenti, M.; Cervera, M.; Dialami, N.; Seret, A. Computational modeling and sub-grid scale stabilization of incompressibility and convection in the numerical simulation of friction stir welding processes. *Arch. Comput. Methods Eng.* **2014**, *21*, 3–37. [CrossRef]
4. Dialami, N.; Chiumenti, M.; Cervera, M.; Agelet de Saracibar, C. An apropos kinematic framework for the numerical modeling of friction stir welding. *Comput. Struct.* **2013**, *117*, 48–57. [CrossRef]
5. Dialami, N.; Chiumenti, M.; Cervera, M.; Agelet de Saracibar, C.; Ponthot, J.P.; Bussetta, P. Numerical Simulation and Visualization of Material Flow in Friction Stir Welding via Particle Tracing. In *Numerical Simulations of Coupled Problems in Engineering*; Idelsohn, S., Ed.; Springer: Cham, Switzerland, 2014; Volume 33, pp. 157–169. ISBN1 978-3-319-06135-1. (Print); ISBN2 978-3-319-06136-8. (Online). [CrossRef]
6. Chiumenti, M.; Cervera, M.; Agelet de Saracibar, C.; Dialami, N. Numerical modeling of friction stir welding processes. *Comput. Methods Appl. Mech. Eng.* **2013**, *254*, 353–369. [CrossRef]

Article

Effect of the Tool Tilt Angle on the Heat Generation and the Material Flow in Friction Stir Welding

Narges Dialami *, Miguel Cervera and Michele Chiumenti

International Center for Numerical Methods in Engineering (CIMNE), Technical University of Catalonia,
Campus Norte UPC, 08034 Barcelona, Spain; Miguel.Cervera@upc.edu (M.C.); Michele@cimne.upc.edu (M.C.)
* Correspondence: Narges@cimne.upc.edu; Tel.: +34-93-401-6529

Received: 5 December 2018; Accepted: 24 December 2018; Published: 29 December 2018

Abstract: This work studies the effect of the tool tilt angle on the generated heat and the material flow in the work pieces joint by Friction Stir Welding (FSW). An apropos kinematic framework together with a two-stage speed-up strategy is adopted to simulate the FSW problem. The effect of tilt angle on the FSWelds is modeled through the contact condition by modifying an enhanced friction model. A rotated friction shear stress is proposed, the angle of rotation depending on the process parameters and the tilt angle. The proposed rotation angle is calibrated by the experimental data provided for a tilt angle 2.5°. The differences of generated heat and material flow for the cases of tool with tilt angle of 0° and 2.5° are discussed. It is concluded that due to the higher temperature, softer material and greater frictional force in the trailing side of the tool, the material flow in the rear side of the FSW tool with the title angle is considerably enhanced, which assists to prevent the generation of defect.

Keywords: FSW; tilt angle; friction; material flow

---

1. Introduction

Friction Stir Welding (FSW) uses a tool with a high rotating speed which moves forward between the pieces to be joined and generates heat. The main function of the tool (consisting of pin and shoulder) is to mix the work piece material and to generate heat by friction. The final properties of friction stir welds depend on factors such as the process parameters (advancing and rotating speed), the tool design and the tool tilt angle [1–3]. In previous works, the authors have studied the effects of the tool velocity [4] and the tool design [5]. In this work, the effect of the tool tilt angle is addressed.

Figure 1 presents a cross-sectional view of an (exaggeratedly) tilted tool inside the work piece. Typical tilt angles used in practice are between 0° and 3°, where a zero value signifies that the tool is perpendicular to the work piece. The tool tilt angle affects the material flow during the weld and thus the heat generation. In FSW, the heat is generated by friction and plastic dissipation. As the mechanical properties are notably temperature-dependent, material flow and heat generation are dependent on each other, making FSW a strongly coupled thermo-mechanical problem. The tool tilt angle has a fundamental importance for the weld quality in FSW. On the one hand, a non-zero tilt angle ensures the contact among the tool shoulder and the work piece; moreover, it facilitates the flow of the material around the tool. On the other hand, an inadequately large tilt angle raises the pin from the weld root, resulting in damaged welds. Consequently, it is essential to properly choose the tool tilt angle. An optimal tool tilt angle guarantees that the tool shoulder imprisons the deformed material and transports it proficiently from the front edge to the rear side of the pin [6].

The tilt angle of the tool and its noticeable effect on the final post-weld quality has been studied by several investigators [7–9]. These studies show that the tool tilt angle has a significant effect on the formation of defects during the weld. The optimal tool tilt angle facilitates the material flow around the tool and avoids the formation of defects in the weld zone. Several experimental tests have to be performed to obtain the optimal tilt angle. However, the fundamental mechanism of the tilt effect on heat generation and material flow is yet to be understood.

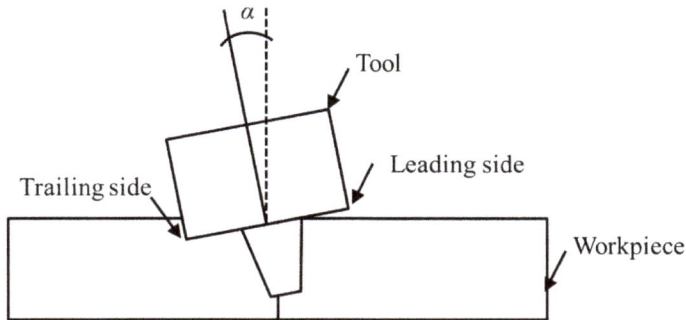

Figure 1. Cross-sectional view of an (exaggeratedly) tilted tool inside the work piece.

Reshad et al. [10] study the effects of the tool tilt angle on FSW of pure titanium. They considered several test cases where the tool tilt angle is varied and the effect of this variation on the post-weld properties is examined. They obtained $1°$ as the best tilt angle for a defect-free welding with high mechanical properties.

Banik et al. [11] examine weld qualities of FSW AA6061-T6 from the point of view of the final mechanical properties of the work piece by changing the tool tilt angles for taper featureless and taper threaded tools. They observe that an increase of the tool tilt angle increases the forces and the torque at the tool/work piece interface.

Elyasi et al. [12] study the effect of the tilt angle on FSW of dissimilar alloys (aluminum to steel). Tilt angles of $1°$, $2°$ and $3°$ are chosen. They observe that a larger tilt angle increases the axial force and the interaction between aluminum and steel.

Hamid and Roslee [13] investigate the tilt angle effect on microstructural and mechanical characteristics of FSWelded dissimilar aluminum alloys. They observe that the tilt angle affects the mechanical properties of the FSW joints considerably. Microstructure of the weld also changes significantly by varying the tilt angle, specifically in the area of weld nugget and heat affected zone.

Meshram and Reddy [14] study the role of the tilt angle on defects generation and material flow in FSW. They observe that the variation of tool tilt angle changes the thermo-mechanical results during FSW and therefore alters the material flow in the weld and controls the weld defects.

In spite of the importance of the effect that the tool tilt angle has on the final quality of the welded work piece, there are only a few computational studies of this phenomenon. Numerical simulations may provide detailed knowledge of the process from both thermal and mechanical point of views.

Long et al. [15] present a 3D thermo-mechanical model with a non-zero tilt angle and study its effect on the final joint. They use DEFORM-3D to simulate the FSW process in a Lagrangian framework. In their work, the tilt angle is considered inside of the geometrical model. They test two cases of $0°$ and $2°$ tilt angle. Wormhole defects are observed in case of $0°$, while the weld in case of $2°$ is defect free.

Chauhan et al. [16] investigate the effect of three tilt angles ($0°$, $1°$ and $2°$) on the formation of defects in FSW applying a Coupled Eulerian and Lagrangian (CEL) method. They use ABAQUS/Explicit to model FSW process with a cylindrical pin. In order to avoid the serious mesh distortion encountered when modeling FSW, the work piece is defined as a Eulerian body. Their model predicts that a tilt angle of $2°$ produces a defect free weld.

Aghajani Derazkola and Simchi [17] present experimental and numerical analysis of friction stir welding of poly (methyl methacrylate) work pieces. They study the effect of process parameters such as tilt angle to define the appropriate conditions for seeking defect-free joints. They observe that the tool tilt angle affects the material flow around the tool. The applied downward forging force needs to be increased for increasing tilt angles, and this results in more frictional heat generation.

There are forecast models based on Artificial Neural Network (ANN). In these cases, both experimental and numerical data are collected to correlate process parameters with technical features of the welded joint. Hamilton et al. [18] integrate differential scanning calorimetry curves for 2017A and 7075 in an existing computational model of the FSW process for heat generation and material flow to create the phase transformations maps occurring in the weld zone. The tool tilt angle of 1.5 is considered during the process. They observe that close to the weld tool, the processing temperatures dissolve fully the equilibrium phase in 7075 and partially in 2017A. Casalino et al. [19] implement ANN in order to investigate the effects of process parameters on the laser welding process quality. Using statistical estimation, the relevance of the process parameters with the weld geometry is studied. It is demonstrated that ANN modeling is beneficial for optimizing the quality of manufacturing processes. Pathak and Jaiswal [20] provide a review on the applications of ANN in FSW. They consider the tilt angle as one of the controlling factor. They conclude that ANN results are matching with the experimental data.

From the previous works, it can be concluded that the tilt angle has a significant effect on the heat generation and material flow and is a controlling parameter to produce a defect free joint.

In previous works devoted to the numerical modeling of the effect of the tilt angle, this angle was considered in the geometrical setting, but not in the contact condition at the tool/work piece interface. Reference [21] is one of the few works, both experimental and numerical, to address heat and mass transfer due to the tilt angle. They use an Eulerian framework for an axisymmetric pin and an incomplete contact boundary condition that applies frictional tangential force on a contact area defined based on the tilt angle (α) and an in plane rotating angle (β) of the contact area. From the experimental evidence, they conclude that this in plane rotating angle is 45° and they use it in the numerical analysis.

In this work, we address the numerical analysis of the effect of the tool tilt angle on FSW from the computational approach developed previously by the authors [5]. It allows obtaining the steady state rapidly at the speed-up phase of the simulation. This is followed by a periodic stage simulation, assuming the first stage as the initial condition. An apropos kinematic system is used by mixing Arbitrary Lagrangian Eulerian (ALE), Eulerian and Lagrangian schemes for different areas of the computational model. The framework can accommodate any pin shapes.

The influence of the tilting is to be represented by the enhanced friction model accounting for the effect of non-uniform pressure distribution under the tool and tilting. The friction model is modified by introducing an in plane rotating angle (β) which depends on the tool tilt angle (α) and the advancing and rotating velocities. In the current study, this parameter is calibrated from the temperature field obtained experimentally for the tilt angle 2.5° presented in reference [21]. Alternatively, the rotating angle β can be obtained experimentally from the relationship between the longitudinal and the transversal forces exerted on the tool.

The outline of this paper is as follows. In Section 2, the general solution strategy used in this work is explained. In Section 3 the modified friction model considering the effect of the tilt angle is presented and discussed. The last section is devoted to the analysis of tool tilt angle effect on the thermo-mechanical behavior in FSW. Mechanical results including the material flow are presented and compared for the no tilt ($\alpha = 0°$, $\beta = 0°$) and with tilt ($\alpha = 2.5°$, $\beta = 25°$) cases. Lastly, some conclusions are drawn.

2. Solution Strategy

The simulation of FSW can be performed in different kinematic frameworks: Lagrangian, Eulerian and ALE.

In a Lagrangian framework, material moves together with the reference system. Therefore, the material flow during the weld is the direct solution of the problem. However, due to the large material deformation in the stir zone of FSW, the mesh used in this area requires continuous re-meshing during the simulation. Re-meshing introduces a significant computational overhead and re-interpolation errors. Thus, the application of other kinematic frameworks is more attractive.

In an Eulerian framework the movement of the material is defined on a fixed configuration. Therefore, no re-meshing is needed. This framework presents limitations when non axisymmetric tool pin shapes are modeled. In these cases, the boundaries of the model are constantly changing by the rotation of the tool pin. Thus re-characterization of the integration domain at every time step of the analysis is indispensable.

The alternative to Lagrangian and Eulerian approaches is an ALE framework where the reference system is not fixed and allowed to move independently from the material movement. An ALE framework permits to treat arbitrary pin geometries and re-meshing can be avoided using a mesh around the tool that rotates rigidly together with the tool.

In a Lagrangian framework, the tool tilt angle can be directly included in the geometrical modeling. In the Eulerian and ALE frameworks, the geometrical model can include the tool tilt angle directly only if the pin shape is axisymmetric. Tilted non-axisymmetric pin shapes require specific ALE approaches [22] as the rotation of the tool is not synchronized with the rotation of the mesh around the tool.

In this work a feasible kinematic framework and a two-stage (speed-up and periodic stages) strategy are adopted for the solution of the overall problem [5,23] (Figure 2). The strategy uses a fully coupled thermo-mechanical framework at both stages. The solution of the coupled thermo-mechanical problem is acquired by performing a staggered time-stepping algorithm solving the thermal and mechanical sub-problems sequentially for each time step.

The speed-up stage aims at obtaining the steady state rapidly by modifying the thermal inertia term in the energy balance equation. At this stage an Eulerian formulation is used.

The periodic stage considers the results obtained at the first stage as an initial condition. At this stage an apropos kinematic framework is used [23]. The choice of this framework is for combining the benefits of ALE, Eulerian and Lagrangian formulations by applying them in the stir zone, the remains of the work piece and the pin-tool, respectively.

The main effect of the tilt angle on the process behavior is the heat generation and its influence on the material flow during FSW. As the sources of heat generation in FSW are plastic dissipation and friction, the suitable modification of the friction law is the strategy proposed here in order to include the effect of the tilt angle.

All the implementations used for this work are done in the in-house finite element code COMET [24] developed by the authors. Details on the technical and computational aspects of the formulation are given in the references [5,23,25].

The resulting model incorporates a two-stage strategy that can speed up the transient stage to obtain the periodic stage with 50 times reduced computational costs comparing with the standard models [5]. Moreover, the model is enriched with an enhanced friction model that considers the real process behavior for generating the frictional heat and can consider the effect of the tilt angle in the heat generation and material flow.

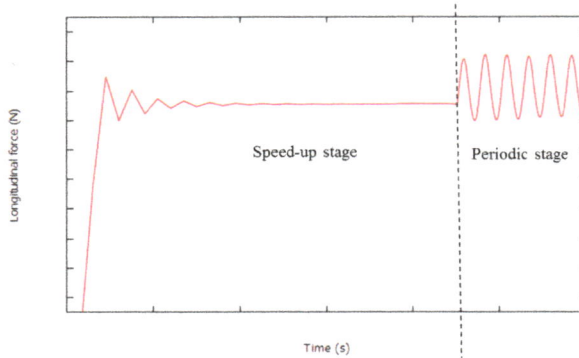

Figure 2. Two-stage strategy concept.

3. Friction Model Including the Tilt Angle

The friction law describes the contact condition at the interface between the tool and the work piece as indicated by their relative sliding velocities. Coulomb's [26–29] and Norton's [22] friction laws are regularly utilized in FSW simulation.

In the previous work of the authors [30], a modified Norton's law is proposed considering the non-uniform pressure distribution that is generally found under the tool during FSW. The enhanced friction model defines the friction shear stress at each point at the contact surface as

$$\boldsymbol{\tau}_T = 0.5 \left(\tau_{max} + \tau_{min} + (\tau_{max} - \tau_{min}) \tanh \frac{x}{R/6} \right) \|\Delta \mathbf{v}_T\|^{q-1} \Delta \mathbf{v}_T, \tag{1}$$

where $\boldsymbol{\tau}_T$ is the friction shear stress, $\Delta \mathbf{v}_T$ is the sliding velocity, $0 \leq q \leq 1$ is the sensitivity parameter, x is the location of each point at the tool/work piece contact surface relative to the rotation axis projected on the welding direction and R is the shoulder radius. τ_{max} and τ_{min} are the maximum and the minimum friction tractions. Figure 3 presents a schematic view of this distribution of friction traction where the average value of the friction is at the center of the tool ($x = 0$). Note that the τ_{max} and τ_{min} values are attained at the leading and trailing edges of the shoulder.

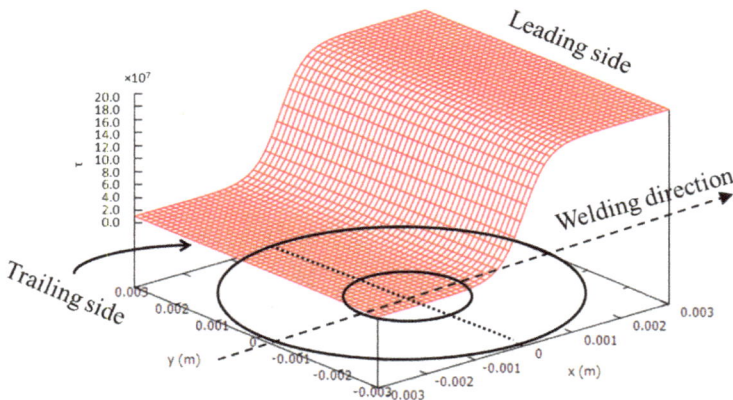

Figure 3. Friction shear traction distribution under the tool (0° tilt angle).

This friction law in Equation (1) does not take into account the effect of the tilt angle. In case of having a tilt angle α (backwards), as the FSW tool advances in the weld direction, the contact surface that has a maximum friction value in front of the tool and minimum friction in the rear side rotates a certain angle β (counterclockwise) around the (counterclockwise) rotating axis due to the tilt influence [21]. This is detected from the experimental evidence in the reference [21] where the effect of tilting appears on the rotation of the contact print. Figure 4 shows schematically what is observed in the experiments. Therefore, the maximum friction is not at the front side but it is rotated by an angle β. The tilting of the tool results in the subsequent rotation of the average friction line. Figure 5 shows schematically how the distribution of the friction under the shoulder is affected by the tool tilt. The x-axis is along the welding direction with or without tilt angle; it is perpendicular to the average friction line (y-axis) when no tilt angle exists. For a tilt angle α, the average friction line (y'-axis) is rotated an angle β in the horizontal plane.

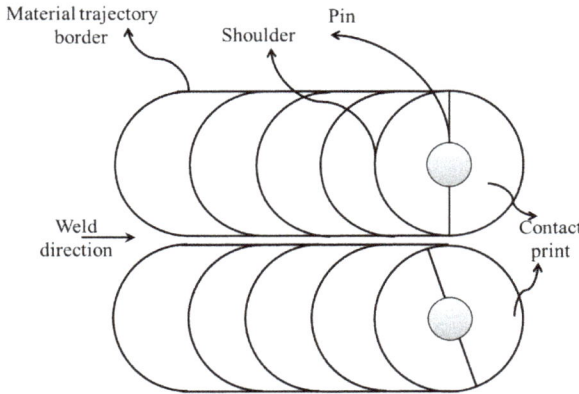

Figure 4. Schematic contact print observed in the experiment of FSW. Without tilt angle (**top**); with tilt angle (**bottom**).

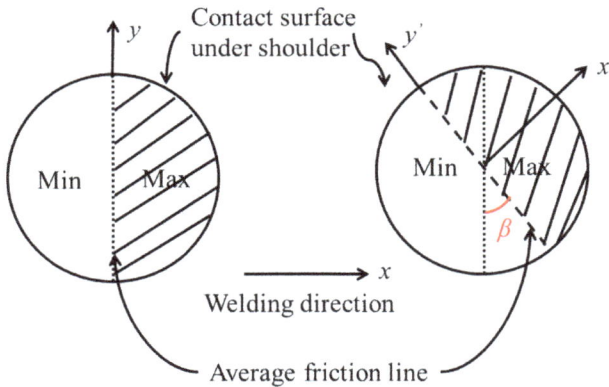

Figure 5. Contact condition under the (counterclockwise) rotating tool. Without tilt angle (**left**); with tilt angle (**right**).

The angle β of the friction shear stress distribution depends on the welding parameters such as tilt angle α, rotating speed ω and advancing speed v_{adv}:

$$\beta = f(\alpha, \omega, v_{adv}) \tag{2}$$

In this work, angle β is obtained by calibration from the temperature field of the experiment presented in [21]. However, the rotating angle β can be obtained experimentally from the relationship between the longitudinal (F_x) and the transversal (F_y) forces exerted on the tool.

$$\beta = \tan^{-1}\left(\frac{F_x}{F_y}\right) \tag{3}$$

The detailed numerical investigation and its experimental validation of the dependence of angle β on the process parameters are out of the scope of this work. We will focus on the influence of angle β on the thermo-mechanical behavior of FSW, by comparing two cases: without tilt ($\alpha = 0°$, $\beta = 0°$) and with tilt ($\alpha = 2.5°$, $\beta = 25°$).

Considering the effect of the tilt and the rotation of the contact shear stress between the tool and the work piece, the reference axes x and y are rotated to the new position x' and y'.

$$\begin{bmatrix} x' \\ y' \end{bmatrix} = \begin{bmatrix} \cos\beta & \sin\beta \\ -\sin\beta & \cos\beta \end{bmatrix} \begin{bmatrix} x \\ y \end{bmatrix}, \tag{4}$$

where β is the rotating angle of the contact surface. Therefore, Equation (1) can be rewritten as

$$\tau_T = 0.5\left(\tau_{max} + \tau_{min} + (\tau_{max} - \tau_{min}) \tanh\frac{x\cos\beta + y\sin\beta}{R/6}\right)\|\Delta\mathbf{v}_T\|^{q-1}\Delta\mathbf{v}_T, \tag{5}$$

Figure 6 presents the distribution of friction law in case of having a tilt angle. The average value of the friction is rotated around the center of the tool.

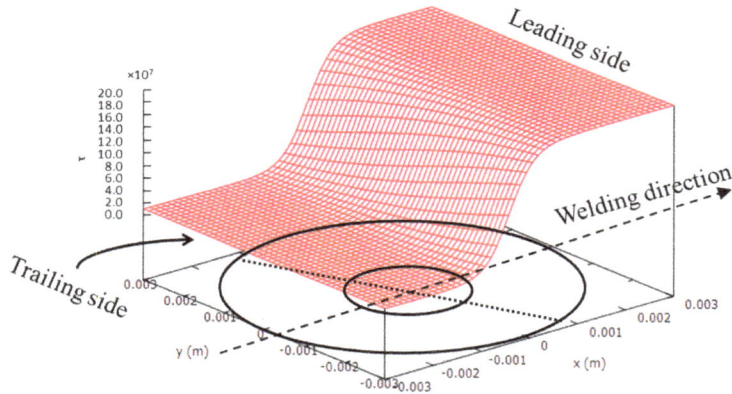

Figure 6. Friction shear traction distribution under the tool (2.5° tilt angle).

4. Analysis of the Effect of the Tilt Angle

In this section, the effect of the tool tilt angle on the thermo-mechanical results of a FSW simulation is studied. The thermal effects are studied through the temperature field. The mechanical effects are analyzed via velocity, stress and strain rate fields and material flow around the tool.

Two cases are considered: 0° and 2.5° tilting angle. The two cases are identical from the point of view of material, processing parameters, geometry and they only differ in tilt angle.

The temperature field in case of having tilt angle obtained from numerical analysis is compared with experiment [21] in order to obtain the corresponding angle of rotation β. Then the thermo-mechanical results in both cases of with and without tilt angle are compared against each other.

The material selected is aluminum alloy AA2024-T4. The chemical composition (wt%) of the aluminum alloy AA2024-T4 is Cu = 4.53, Mg = 1.62, Mn = 0.65, Si = 0.066, Fe = 0.21 and Al = Bal. [31].

The dimension of the work piece is $300 \times 75 \times 5$ mm^3. Figure 7 shows the geometry model including tool, stir zone and the rest of the work piece. The tool has a flat shoulder of 16 mm diameter and a featureless conical pin. The top and bottom diameter of the pin are 6 mm and 4 mm, respectively. The height of the pin is 4.8 mm. Figure 8 shows the dimension of the tool used.

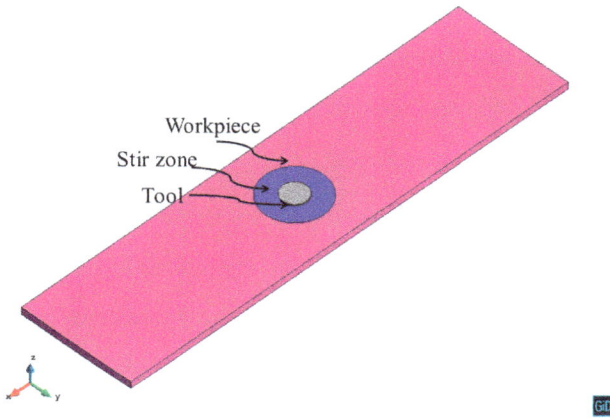

Figure 7. The geometry model.

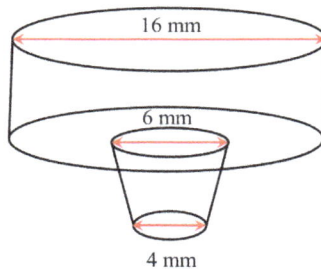

Figure 8. Conical tool geometry.

The rotating and advancing velocities are 800 rpm and 20 mm/min, respectively.

The process parameters are selected as such to compare the numerical results obtained in this work with the experimental data published in [21].

The computational model consists of 380,000 tetrahedral elements and 60,000 points approximately. Figure 9 shows the corresponding mesh used to discretize the model.

Heat generated through both plastic dissipation and frictional contact is considered.

The visco-plastic dissipation (D_{mech}) is defined as

$$D_{mech} = \gamma \mathbf{s} : \dot{\boldsymbol{\varepsilon}}, \tag{6}$$

where γ is the Taylor–Quinney coefficient, $\mathbf{s}$ is the deviatoric stress and $\dot{\varepsilon}$ is the strain rate.

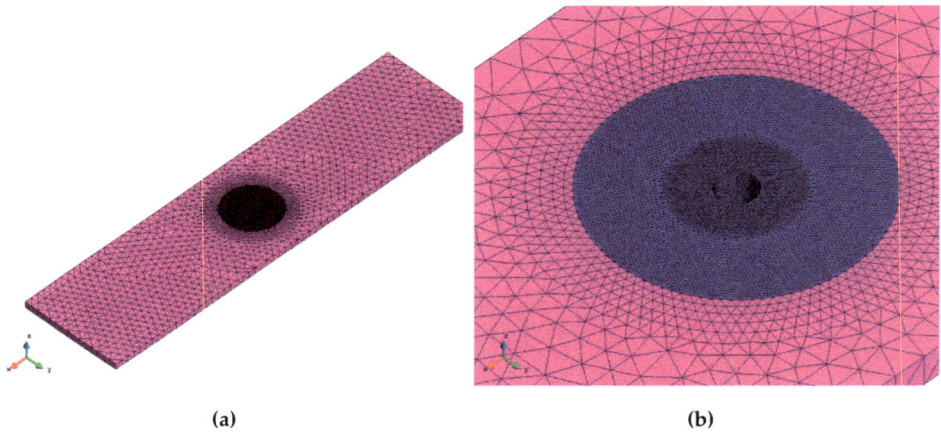

Figure 9. Mesh resolution. (**a**) entire domain; (**b**) zoom on the stir zone.

The analysis considers the minimum and maximum friction tractions as $\tau_{min} = 1.55 \times 10^7$ and $\tau_{max} = 3.1 \times 10^8$ at tool/wor-kpiece contact surface, respectively. They are obtained from the calibration from the temperature field [30]. The in plane rotating angle $\beta = 25°$ is calibrated from the temperature field obtained experimentally for the tilt angle 2.5° and presented in reference [21]. The qualitative mechanical results including the material flow are presented and compared for the cases ($\alpha = 0°$, $\beta = 0°$) and ($\alpha = 2.5°$, $\beta = 25°$).

To verify the choice of the rotating angle $\beta = 25°$, the longitudinal and the transversal forces are evaluated in both cases, without and with tilt angle. In the first case (without tilt angle), the forces are $F_x = 170$ N and $F_y = 28000$ N. Thus according to Equation (3), the in plane rotation angle is $\beta \cong 0°$. In the second case (with tilt angle), the forces are $F_x = 12000$ N and $F_y = 25000$ N. Thus the in plane rotation angle is effectively $\beta = 25°$.

4.1. Thermal Effects

The thermal effects caused by the tool tilt angle are presented in this section in terms of the temperature field at the steady state.

Temperature

Figure 10 shows the computed temperature field for the tilt angles of 0° and 2.5°. The results are also shown on a vertical section at the center of the tool and the leading side in order to see the temperature field on the top surface and within the depth of the work piece.

The difference caused by the tool tilt angle on the temperature distribution can be clearly seen. In both cases, the maximum temperature is on the advancing side (AS) due to the non-uniform distribution of the friction at the contact surface between tool and work piece. The tool tilt angle causes a slight rotation of the temperature field. It increases the temperature in the neighboring zone of the FSW tool in the rear advancing side. This observation is in accordance with the experimental finding in [32].

The difference of the temperature in the advancing (AS) and the retreating sides (RS) diminishes with the distance from the tool center. The computed maximum temperatures are compared with the measured ones presented in [21] for tilt angle 2.5° at different distances from the welding line on both advancing and retreating side, see Figures 11 and 12.

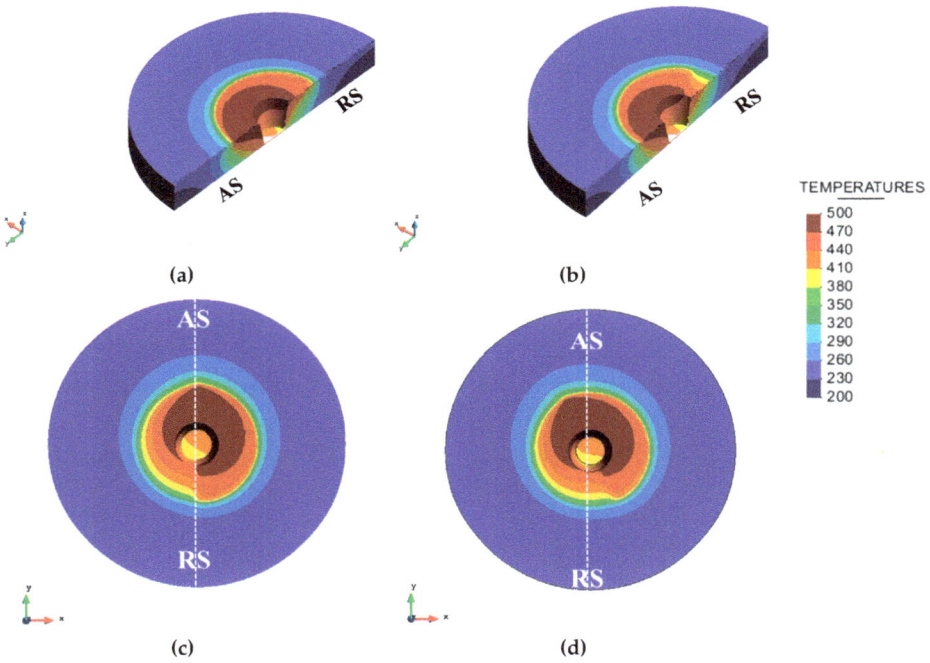

Figure 10. Temperature distribution. (**a,c**) 0° tilt angle; (**b,d**) 2.5° tilt angle.

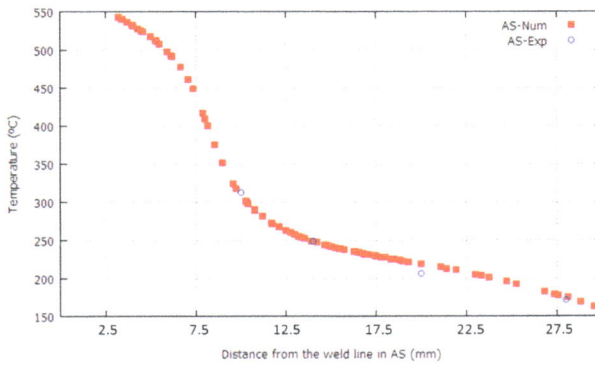

Figure 11. Temperature comparison between numerical results and experimental data on the advancing side (AS) at different distances from the weld line (2.5° tilt angle).

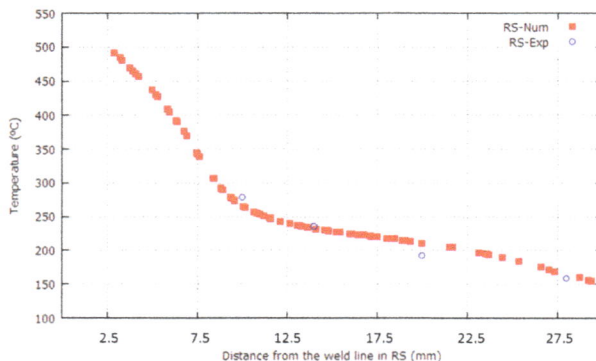

Figure 12. Temperature comparison between numerical results and experimental data on the retreating side (RS) at different distances from the weld line (2.5° tilt angle).

It can be seen from the figures that a good agreement is achieved between numerical and experimental results at both retreating and advancing sides of the FSWelded work piece.

Table 1 compares the differences between the temperatures at retreating and advancing sides obtained from experimental and numerical analyses when a tilt angle of 2.5° is used. This comparison is performed for distances of 10, 14, 20 and 28 mm from the weld line. The agreement between numerical results and experimental data is remarkable.

Table 1. Difference between temperatures on the advancing and retreating sides at different locations from the weld line.

Distance from the Welding Line (mm)	Temperature Difference of AS and RS (Tilt Angle of 2.5°)	
	Experimental Data	**Numerical Analysis**
10	35	36
14	15	15
20	13	11
28	11	10

4.2. Mechanical Effects

The mechanical effects caused by the tool tilt angle are presented in this section in terms of velocity, stresses and strain rate fields and material flow around the tool.

4.2.1. Velocity, Stress and Strain Rate

The computed velocity fields for $\alpha = 0°$ and $\alpha = 2.5°$ under the shoulder are depicted in Figure 13.

The velocity field in case of tilt angle $\alpha = 2.5°$ is rotated $\beta = 25°$ (obtained from the calibration of the rotating angle β from the temperature field presented in [21]). As expected, the maximum velocity is at the border of the shoulder and in case of 0° tilt angle at the leading edge.

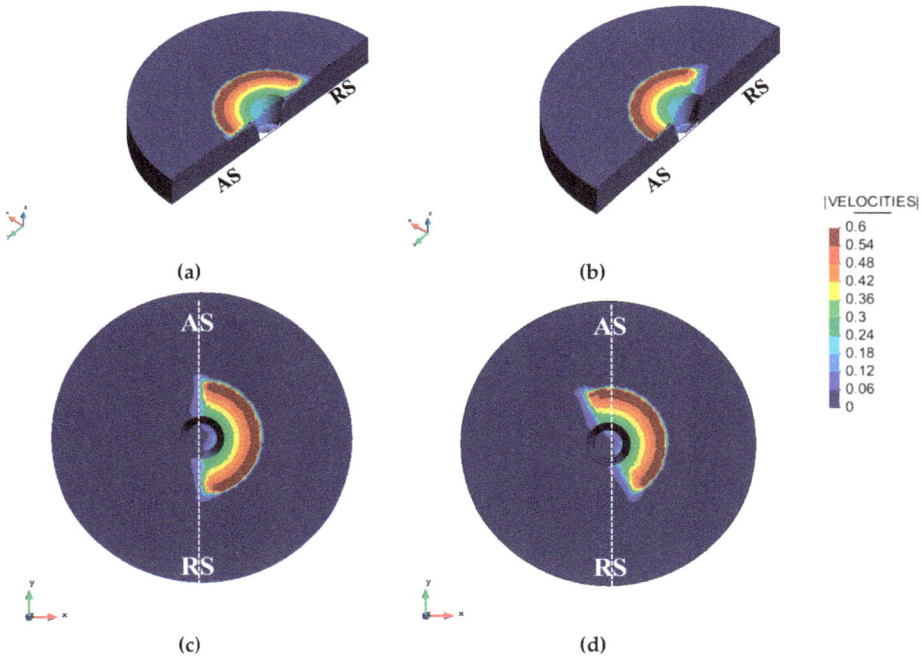

Figure 13. Velocity field. (**a**,**c**) 0° tilt angle (**b**,**d**) 2.5° tilt angle.

The J2 stress distributions under the shoulder in both studied cases are presented in Figure 14. The effect of the tilt angle is the rotation of this distribution and the increase of the stresses in the retreating side of the leading front. As the tool tilt angle increases the temperature on the rear advancing side of the tool (Figure 10), the material flow stress decreases correspondingly in this region and consequent softening of the material facilitates the flow. The computed behavior agrees with the finding in [32].

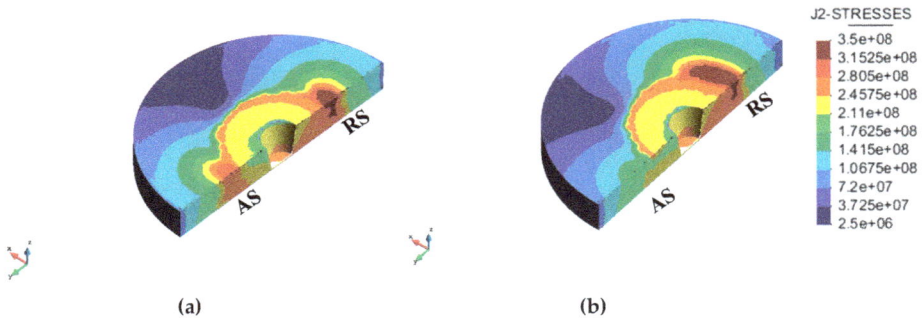

Figure 14. *Cont.*

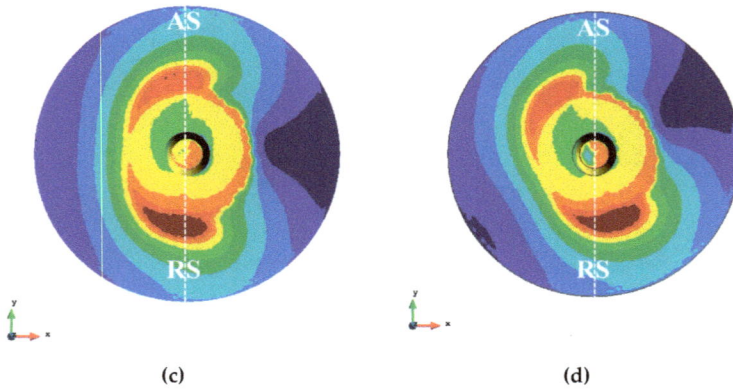

Figure 14. J2 stress field. (**a,c**) 0° tilt angle; (**b,d**) 2.5° tilt angle.

The strain rate distribution is shown in Figure 15 for both tilt angles. As the strain rate defines the stirring action in FSW [33], the distribution of the strain rate under the tool can give an insight to the material stirring. In case of having tilt angle, the stirring effect increases on the rear edge of the tool on the advancing side. Therefore the tilt angle can strengthen the material stirring action at this zone.

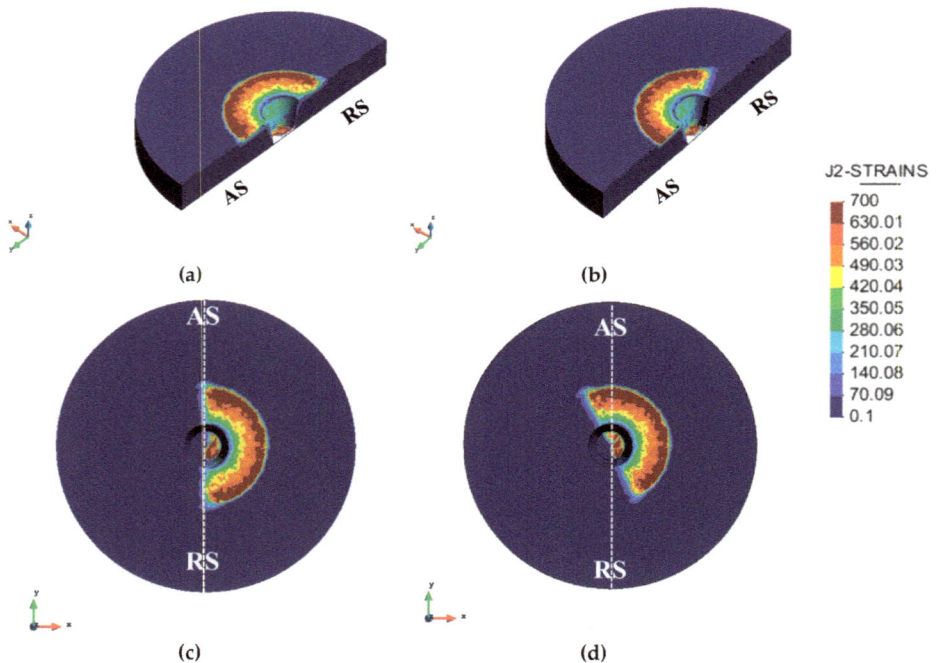

Figure 15. J2 strain rate field. (**a,c**) 0° tilt angle; (**b,d**) 2.5° tilt angle.

4.2.2. Material Flow

In order to visualize the differences on the material flow around the tool, stream lines are shown in Figures 16 and 17 for both tilting cases at different depths (1 mm and 3 mm). In the xy plane, the effect of tilting manifests in the rotation of the streamlines (Figures 16b and 17b).

In the case of no tilt angle and 1 mm depth, the material moves extensively around the tool. It is visible both in the xy plane (Figure 16a) and the xz plane showing that the material passes through all the depth of the tool (Figure 16c).

The material movement at the depth of 1 mm is very much affected by the shoulder movement (Figure 16c,d). Due to the tilt angle effect, material is trapped by the shoulder on the trailing edge (Figure 16d). It can be seen from Figure 16d that tilting induces a considerable accelerating effect behind the pin tool which is difficult to achieve in case of having no tilt angle. The reason for this is that a higher friction force exists on the rear side of the tool at the interface between the tool and the work piece. This stimulating effect of the tool tilt on the material flow helps to avoid the formation of defects in the weld, as low material velocity might lead to defect formation [34].

Further from the shoulder, at the depth of 3mm, the effect of tilting is less evident. In both cases of 0° and 2.5° tilt angles, apart from the rotation of the streamlines, the material movement around the tool is similar (Figure 17a,b). Without tilt angle, the material around the tool goes through all the depth of the pin resembling the case with tilt angle (Figure 17c,d).

It can be seen from Figure 17d that the material flows around the pin and then accumulates behind the tool on the weld.

Due to the higher temperature, softer material and greater frictional force in the trailing side of the tool, the material flow in the rear side of the FSW tool with the title angle is considerably enhanced, which assists to prevent the generation of defect.

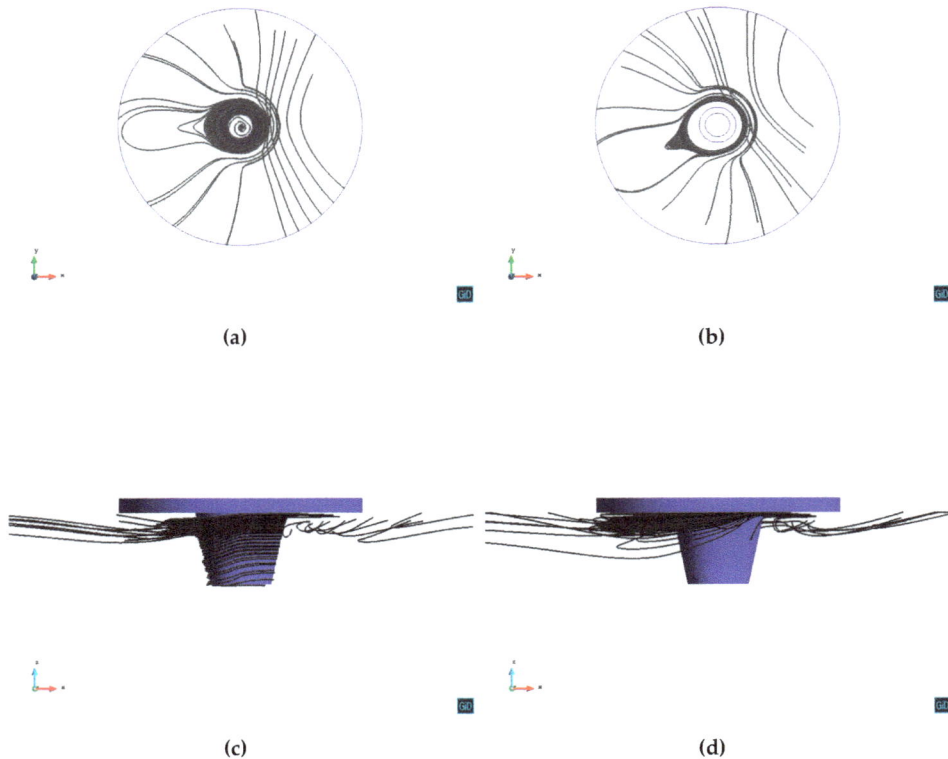

Figure 16. Material flow paths in horizontal and vertical view at depth of 1 mm. (**a**) 0° tilt angle, depth of 1 mm, xy plane; (**b**) 2.5° tilt angle, depth of 1 mm, xy plane; (**c**) 0° tilt angle, depth of 1 mm, xz plane; (**d**) 2.5° tilt angle, depth of 1 mm, xz plane.

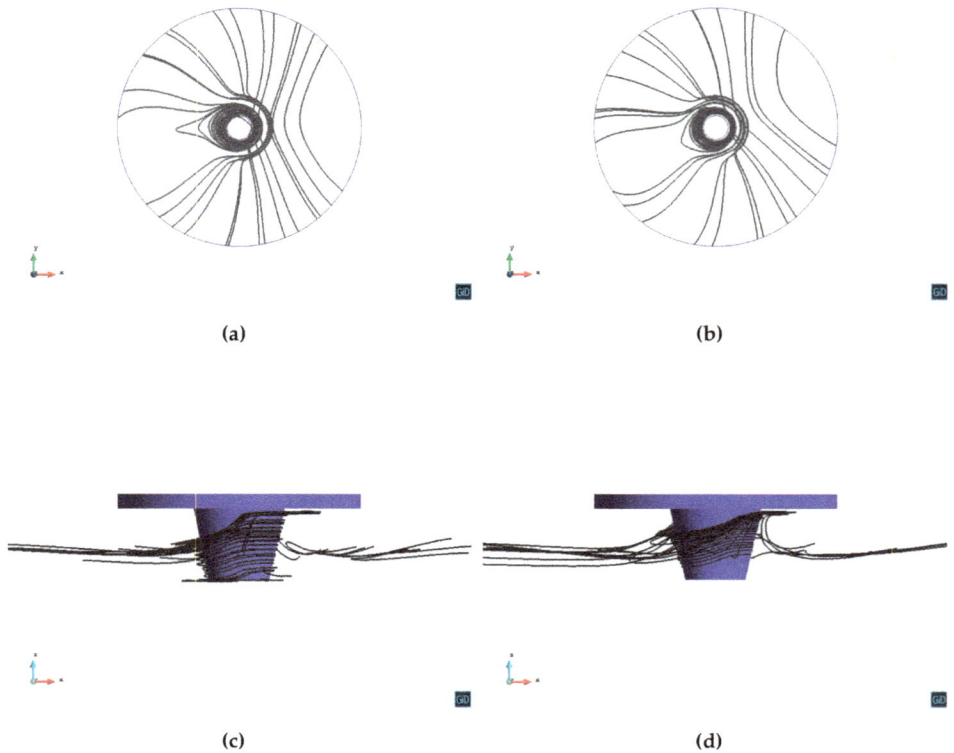

(a)

(b)

(c)

(d)

Figure 17. Material flow paths in horizontal and vertical view at depth of 3 mm. (**a**) 0° tilt angle, depth of 3 mm, xy plane; (**b**) 2.5° tilt angle, depth of 3 mm, xy plane; (**c**) 0° tilt angle, depth of 3 mm, xz plane; (**d**) 2.5° tilt angle, depth of 3 mm, xz plane.

5. Summary and Conclusions

In this work, the effect of tool tilt angle on the thermo-mechanical results (heat generation and material flow) in FSW process is studied. The thermo-mechanical results are presented for two tilt angle cases of $\alpha = 0°$ and $\alpha = 2.5°$.

The friction model is modified by introducing an in plane rotating angle β of the friction shear stress in order to account for the effect of tilting. This rotation angle is calibrated from the temperature field obtained experimentally for the tilt angle 2.5°. The qualitative mechanical results including the material flow are presented and compared for the cases ($\alpha = 0°$, $\beta = 0°$) and ($\alpha = 2.5°$, $\beta = 25°$). It is verified that the rotating angle β can be defined through the relationship between the longitudinal and the transversal forces exerted on the tool.

It is observed that a non-zero tilt angle results in the rotation of the friction distribution under the shoulder. The computed temperature is compared with the experimental data and good agreement is obtained at both advancing and retreating sides. Differences between the temperatures at retreating and advancing sides are also compared with the experimental measurements.

It is observed that the tool tilt angle:

- increases stresses at the leading edge of the tool on the work piece.
- increases the temperature in the neighboring zone of the FSW tool in the rear advancing side.
- decreases the material flow stress in the rear advancing side.
- strengthens the material stirring action at trailing edge on the advancing side.

- facilitates the material flow behind the tool.

To sum up, the effect of the tool tilt angle can be represented by the in plane rotation of the corresponding thermo mechanical fields. Owing to this, the material flow behind the FSW tool is considerably improved contributing to the prevention of the defect formation. Future work will be addressed to extend the experimental evidence in order to correlate the tilt angle and process parameters with obtained in plane rotation of the friction shear traction.

Author Contributions: N.D., M.C. (Miguel Cervera) and M.C. (Michele Chiumenti) developed the numerical model; N.D. implemented the model; N.D. and M.C. (Miguel Cervera) discussed and analyzed the data.

Funding: This research received no external funding.

Conflicts of Interest: The authors declare no conflict of interest.

References

1. Langari, J.; Kolahan, F.; Aliakbari, K. Effect of tool speed on axial force, mechanical properties and weld morphology of friction stir welded joints of a7075-t651. *Int. J. Eng.* **2016**, *29*, 403–410.
2. Mehta, K.P.; Badheka, V.J. Effects of tilt angle on properties of dissimilar friction stir welding copper to aluminum. *Mater. Manuf. Process.* **2016**, *31*, 255–263. [CrossRef]
3. Jaiganesh, V.; Maruthu, B.; Gopinath, E. Optimization of process parameters on friction stir welding of high density polypropylene plate. *Procedia Eng.* **2014**, *97*, 1957–1965. [CrossRef]
4. Dialami, N.; Chiumenti, M.; Cervera, M.; Agelet de Saracibar, C.; Ponthot, J.-P. Numerical simulation and visualization of material flow in friction stir welding via particle tracing. In *Numerical Simulations of Coupled Problems in Engineering*; Springer International Publishing: Cham, Switzerland, 2014; pp. 157–169.
5. Dialami, N.; Chiumenti, M.; Cervera, M.; de Saracibar, C.A. A fast and accurate two-stage strategy to evaluate the effect of the pin tool profile on metal flow, torque and forces during friction stir welding. *Int. J. Mech. Sci.* **2017**, *122*, 215–227. [CrossRef]
6. Chien, C.H.; Lin, W.B.; Chen, T. Optimal FSW process parameters for aluminum alloys AA5083. *J. Chin. Inst. Eng.* **2011**, *34*, 99–105. [CrossRef]
7. Arici, A.; Selale, S. Effects of tool tilt angle on tensile strength and fracture locations of friction stir welding of polyethylene. *Sci. Technol. Weld. Join.* **2007**, *12*, 536–539. [CrossRef]
8. Payganeh, G.H.; Arab, N.B.M.; Asl, Y.D.; Ghasemi, F.A.; Boroujeni, M.S. Effects of friction stir welding rocess parameters on appearance and strength of polypropylene composite welds. *Int. J. Phys. Sci.* **2011**, *6*, 595–601.
9. Shazly, M.; El-raey, M. Friction stir welding of polycarbonate sheets. In *Characterization of Minerals, Metals, and Materials*; Wiley: Hoboken, NJ, USA, 2014; pp. 555–563.
10. Reshad Seighalani, K.; Besharati Givi, M.K.; Nasiri, A.M.; Bahemmat, P. Investigations on the Effects of the Tool Material, Geometry, and Tilt Angle on Friction Stir Welding of Pure Titanium. *J. Mater. Eng. Perform.* **2010**, *19*, 955. [CrossRef]
11. Banik, A.; Roy, B.S.; Barma, J.D.; Saha, S.C. An experimental investigation of torque and force generation forvarying tool tilt angles and their effects on microstructure and mechanical properties: Friction stir welding of AA 6061-T6. *J. Manuf. Process.* **2018**, *31*, 395. [CrossRef]
12. Elyasi, M.; Aghajani Derazkola, H.; Hosseinzadeh, M. Investigations of tool tilt angle on properties friction stir welding of A441 AISI to AA1100 aluminium. *Proc. Inst. Mech. Eng. Part B J. Eng. Manuf.* **2016**, *230*, 1234–1241. [CrossRef]
13. Hamid, H.A.D.; Roslee, A.A. Study the Role of Friction Stir Welding Tilt Angle on Microstructure and Hardness. *Appl. Mech. Mater.* **2015**, *799–800*, 434–438. [CrossRef]
14. Meshram, S.D.; Reddy, G.M. Influence of Tool Tilt Angle on Material Flow and Defect Generation in Friction Stir Welding of AA2219. *Def. Sci. J.* **2018**, *68*, 512–518.
15. Long, L.; Chen, G.; Zhang, S.; Liu, T.; Shi, Q. Finite-element analysis of the tool tilt angle effect on the formation of friction stir welds. *J. Manuf. Process.* **2017**, *30*, 562–569. [CrossRef]
16. Chauhan, P.; Jain, R.; Pal, S.K.; Singh, S.B. Modeling of defects in friction stir welding using coupled Eulerian and Lagrangian method. *J. Manuf. Process.* **2018**, *34*, 158–166. [CrossRef]

17. Aghajani Derazkola, H.; Simchi, A. Experimental and thermomechanical analysis of friction stir welding of poly(methyl methacrylate) sheets. *Sci. Technol. Weld. Join.* **2018**, *23*, 209–218. [CrossRef]

18. Hamilton, C.; Dymek, S.; Kopyscianski, M.; Weglowska, A.; Pietras, A. Numerically Based Phase Transformation Maps for Dissimilar Aluminum Alloys Joined by Friction Stir-Welding. *Metals* **2018**, *8*, 324. [CrossRef]

19. Casalino, G.; Facchini, F.; Mortello, M.; Mummolo, G. ANN modelling to optimize manufacturing processes: The case of laser welding, IFAC Proceedings. *IFAC-PapersOnline* **2016**, *49*, 378–383. [CrossRef]

20. Pathak, M.; Jaiswal, D. Application of Artificial Neural Network in Friction Stir Welding: A Review. *Int. J. Technol. Explor. Learn.* **2014**, *3*, 513–517.

21. Zhang, S.; Shi, Q.; Liu, Q.; Xie, R.; Zhang, G.; Chen, G. Effects of tool tilt angle on the in-process heat transfer and mass transfer during friction stir welding. *Int. J. Heat Mass Transf.* **2018**, *125*, 32–42. [CrossRef]

22. Guerdoux, S. Numerical Simulation of the Friction Stir Welding Process. Ph.D. Thesis, École Nationale Supérieure des Mines de Paris, Paris, France, 2007.

23. Dialami, N.; Chiumenti, M.; Cervera, M.; de Saracibar, C.A. An apropos kinematic framework for the numerical modeling of friction stir welding. *Comput. Struct.* **2013**, *117*, 48–57. [CrossRef]

24. Cervera, M.; Agelet de Saracibar, C.; Chiumenti, M. COMET: Coupled Mechanical and Thermal Analysis, Data Input Manual, Version 5.0, Technical Report IT-308. 2002. Available online: http://www.cimne.upc.es (accessed on June 2002).

25. Dialami, N.; Cervera, M.; Chiumenti, M.; Agelet de Saracibar, C. Local-global strategy for the prediction of residual stresses in FSW processes. *Int. J. Adv. Manuf. Technol.* **2016**, *88*, 3099–3111. [CrossRef]

26. Veljic, D.M.; Rakin, M.P.; Perovic, M.M. Heat generation during plunge stage in friction stir welding. *Therm. Sci.* **2013**, *17*, 489–496. [CrossRef]

27. Zhang, Z.; Zhang, H.W. Effect of contact model on numerical simulation of friction stir welding. *Acta Metall. Sin.* **2008**, *44*, 85–90.

28. Chao, Y.J.; Qi, X.; Tang, W. Heat transfer in friction stir welding: Experimental and numerical studies. *J. Manuf. Sci. Eng.* **2003**, *125*, 138–145. [CrossRef]

29. Schmidt, H.; Hattel, J. A local model for the thermomechanical conditions in friction stir welding. *Model. Simul. Mater. Sci. Eng.* **2005**, *13*, 77–93. [CrossRef]

30. Dialami, N.; Chiumenti, M.; Cervera, M.; Segatori, A.; Osikowicz, W. Enhanced friction model for Friction Stir Welding (FSW) analysis: Simulation and experimental validation. *Int. J. Mech. Sci.* **2017**, *133*, 555–567. [CrossRef]

31. Jung, J.; Oak, J.-J.; Kim, Y.-H.; Cho, Y.J.; Park, Y.H. Wear Behaviors of Pure Aluminum and Extruded Aluminum Alloy (AA2024-T4) Under Variable Vertical Loads and Linear Speeds. *Met. Mater. Int.* **2017**, *23*, 1097–1105. [CrossRef]

32. Sheppard, T.; Wright, D.S. Determination of flow-stress constitutive equation for aluminum-alloys at elevated-temperatures. *Met. Technol.* **1979**, *6*, 215–223. [CrossRef]

33. Pashazadeh, H.; Teimournezhad, J.; Masoumi, A. Numerical investigation on the mechanical, thermal, metallurgical and material flow characteristics in friction stir welding of copper sheets with experimental verification. *Mater. Des.* **2016**, *55*, 619–632. [CrossRef]

34. Zhu, Y.; Chen, G.; Chen, Q.; Zhang, G.; Shi, Q. Simulation of material plastic flow driven by non-uniform friction force during friction stir welding and related defect prediction. *Mater. Des.* **2016**, *108*, 400–410. [CrossRef]

Article

Influences of Pin Shape on a High Rotation Speed Friction Stir Welding Joint of a 6061-T6 Aluminum Alloy Sheet

Yang Zhou [1,2], Shujin Chen [2,*], Jiayou Wang [2], Penghao Wang [2] and Jingyu Xia [2]

[1] School of Computer Science and Engineering, Jiangsu University of Science and Technology, Zhenjiang 212003, China; zhouy_just@163.com
[2] School of Material Science and Engineering, Jiangsu University of Science and Technology, Zhenjiang 212003, China; zj_jiayouw@126.com (J.W.); penghao_just@126.com (P.W.); wuyunkaigxy@163.com (J.X.)
* Correspondence: 200800002666@just.edu.cn; Tel.:+86-511-84401185

Received: 30 October 2018; Accepted: 21 November 2018; Published: 24 November 2018

Abstract: In order to explore the influences of different pins on the weld based on the specialty of the aluminium alloy sheet welding, three kinds of pins were chosen to perform high rotation speed friction stir welding on a 1 mm thick 6061-T6 aluminium alloy in this study. The microstructure and mechanical properties of the joints were analysed at the same time. When the rotation speed was 11,000 rpm and the welding speed was 300 mm/min, more sufficient stirring and a better joint (the tensile strength reaches 87.2% of the base metal) can be obtained with the pin design of a quadrangular frustum pyramid. The pattern of the weld cross section was a "flat T" and no obvious "S curve" was found in nugget zone (NZ). Heat affected zone (HAZ) and thermo-mechanically affected zone (TMAZ) were also narrow. The results demonstrate that the proportion of low angle boundaries in each area of the weld is lower than that of traditional Friction Stir Welding (FSW). The grain size of NZ is significantly refined and the proportion of low angle boundaries is only 20.1%, which have improved the welding quality.

Keywords: high rotation speed friction stir welding; pin shapes; grain orientation

1. Introduction

Due to its low density and high strength, the aluminum alloy has been widely used in aerospace, automobile, machinery manufacturing, shipping, and chemical industries [1]. Steel is replaced by a high-strength aluminum alloy sheet in order to conserve energy by reducing the vehicle weight, especially in the automobile manufacturing industry [2,3].

Nevertheless, defects and deformation appear during the aluminum alloy sheet welding process as results of non-uniform heating, inappropriate welding parameters, etc. Furthermore, certain defects in the fusion welding process, such as pores or cracks, are attributed to the limitation of the weld ability of the aluminum alloy [4]. However, Friction Stir Welding (FSW) can achieve solid-state welding without filler materials, which effectively avoids cracks and porosity defects [5–10]. Scialpi et al. successfully conducted ultra-micro-friction stir welding on 0.8 mm 2024-T3 and 6082-T6 sheets, and analyzed the mechanical properties [11]. Tong et al. tested traditional FSW on a 1 mm aluminum alloy sheet [12]. The range of welding parameters and relevant mechanical properties were studied in existing work, which mainly focused on traditional FSW with a rotation speed lower than 1000 rpm [13,14].

Recent research reported that the rotation speed can reach 10 times or more that of traditional FSW [15]. For the higher welding speed and the smaller welding deformation, high rotation FSW is more suitable for aluminum alloy sheet welding [16,17]. Additionally, the technique is expected to be used for robotic welding because of its lower axial pressure [18]. Given the value of application, it merits further study. However, there are scarce research studies examining the joints of high rotation speed friction stir welding (HSFSW).

It is well-known that the pin shape of the tool impacts the FSW process. Additionally, the geometric optimization of the pin shape influences the welding quality [19]. Elangovana et al. pointed out that the tool geometry is a predominant factor determining the weld forming, localized heating, and stirring action [20]. At the same time, the plastic metal flow behavior is mainly influenced by the pin profile, pin dimensions, and FSW process parameters [21]. Compared with the traditional FSW, the smaller size tool is needed and less material is involved in the plastic metal flow in the HSFSW welding process [22,23]. However, few research studies have studied the effects of the pin on the weld microstructure under the high rotation speed condition [24,25]. Therefore, it is necessary to examine the impacts of the pin shape on the welding quality and the microstructure of the joint.

It is important to design a reasonable pin so as to stir these few plastic metals effectively. In this study, three kinds of pin shape were chosen to perform HSFSW on a 1mm thick 6061-T6 aluminum alloy. The hardness and microstructure of the weld cross-section were analyzed. Furthermore, the Electron Backscattered Diffraction system (EBSD) samples were also prepared to reveal the microstructure and mechanical properties under the condition of high rotation speed.

2. Experiment Materials and Methods

The high rotation speed FSW machine used in this study is shown in Figure 1. The FSW tools are all made of hot-work abrasives steel. The diameter of the tool shoulder is 7 mm, but the shapes of the pins are different (Figure 2). FSW tools with different pins are denoted as S1, S2 and S3. As shown in Figure 2, S1, S2 and S3 are a quadrangular prism, quadrangular frustum pyramid, and frustum, respectively. The length of these pins is 0.9 mm.

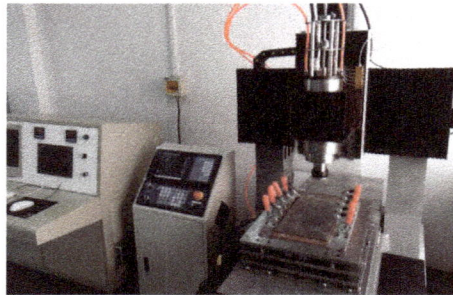

Figure 1. High rotation speed friction stir welding (HSFSW) machine.

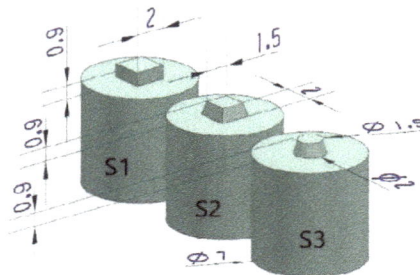

Figure 2. The size of pins, *l*/mm.

The base metal selected is a 6061-T6 aluminum sheet (150 mm × 80 mm × 1 mm) with a tensile strength of 304 MPa, good ductility, corrosion resistance, and no stress corrosion cracking tendency during the welding process [26]. Its composition is shown in Table 1.

Table 1. Composition of 6061-T6 aluminum.

Chemical Composition (mass%)								
Cu	Si	Fe	Mn	Mg	Zn	Cr	Ti	Al
0.15–0.4	0.4–0.8	0.7	0.15	0.8–1.2	0.25	0.04–0.35	0.15	margin

During the prewelding process, the workpieces should be rigidly fixed to the worktable. The butt weld configuration is used in the experiments. The rotation speed selected during the welding process is 11,000 rpm, and the welding speed varies from 200 mm/min to 500 mm/min. In addition, the position control is selected for each welding experiment, and the plunge depth of the shoulder is kept at 0.05 mm.

3. Experimental Results

3.1. The Visual Testing

The weld surface is shown in Table 2 and Figure 3. As we can see from Table 2, pin shape and travel speed (V) can affect weld surface quality, even though the rotation speed remains the same. When the travel speed is 200 mm/min, the weld surface is poor, regardless of pin shape. In the case of S1, a groove appears along the weld seam and seriously flashes on both sides of the weld. In the case of S2, the surface of the weld is rough. In the case of S3, a groove appears along the weld.

Table 2. Weld surface at different travel speeds.

Pin \ V (mm/min)	200	300	400	500
S1	Flash, groove	smooth (W1)	groove	groove
S2	rough	smooth (W2)	smooth	groove
S3	flash	smooth (W3)	smooth	groove

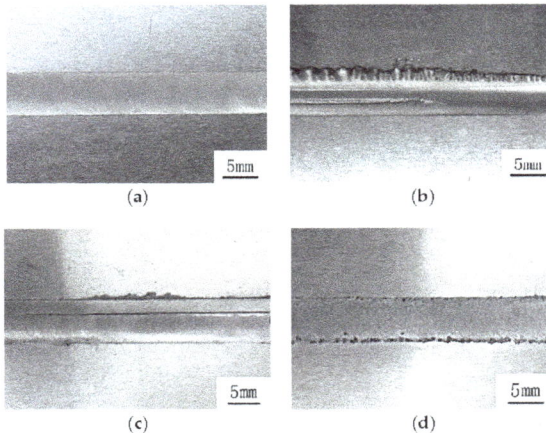

(a)

(b)

(c)

(d)

Figure 3. The surface of the weld: (**a**) smooth (S2, 300 mm/min, 400 mm/min); (**b**) flash and groove (S1, 200 mm/min); (**c**) groove (S3, 500 mm/min); (**d**) rough (S2, 200 mm/min).

When the *V* is 300 mm/min, the soundable appearance of the weld is obtained by using the S1, S2, and S3 tools. But if the *V* reaches 400 mm/min, a smooth surface can only be obtained by using S2 and S3. When the travel speed is 500 mm/min, the surface of all welds has groove defects.

It can be noted that a too low or too high travel speed is not suitable for sheet welding. The lower travel speed leads to the accumulation of heat in the welding area, causing the over plasticization of metal in the welding area and inevitably flashes. The groove will thereupon appear in the case that the plasticized metal is extruding too much. As for the other extreme, the higher travel speed results in inadequate heat input, and then insufficient plasticization makes the metal filling cycle incomplete, leading to groove defects. For the 1 mm thick 6061-T6 aluminum alloy sheet, the welding process window of the S2 FSW tool is wider than that of S1 and S3.

3.2. Axial Force

Due to its lower axial pressure, HSFSW is expected to be used for robotic welding. In the previous experiments, the axial force with different rotation speeds was collected (Figure 4). When the diameter of tool shoulder is 7 mm, the axial force decreases as the rotation speed increases. The average axial force is 1.375 KN while the rotation speed is 11,000 rpm.

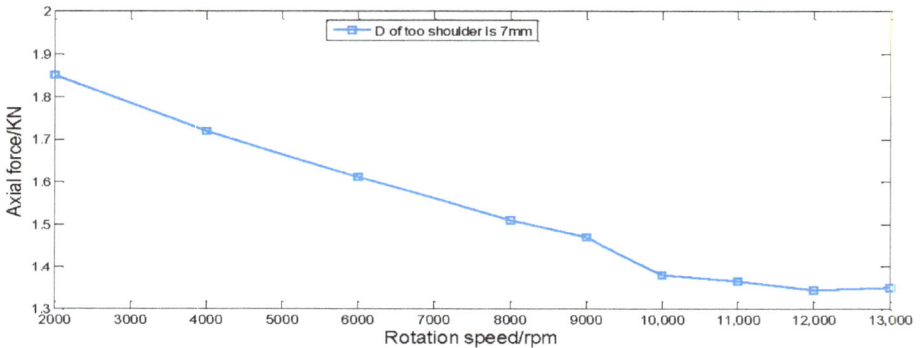

Figure 4. Distribution of axial force with different rotation speeds.

In the experiments for this paper, the axial force with different travel speeds during the whole process was collected (Figure 5). As the diameter of the tool shoulder and the rotation speed are fixed, the axial force during the whole process fluctuates a little, but the average axial force increases while the travel speed also increases.

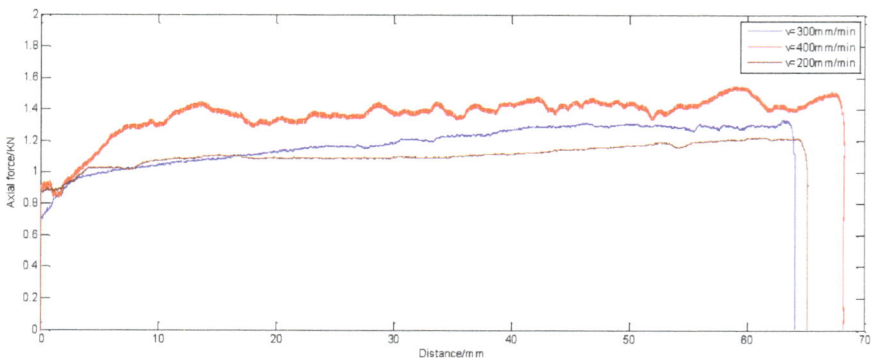

Figure 5. Distribution of axial force during the welding process with different travel speeds.

3.3. Macro Morphology and Micro Hardness

Three tensile specimens (W1, W2, W3) corresponding to S1, S2, and S3 tools were prepared by wire-electrode cutting. As shown in Figure 6, the "flat T" pattern of the cross-sections of the joints W1, W2, and W3 can be observed by a metallographic microscope. Due to the differences between the shapes of the pins, the boundary lines of the joint W1 are approximately perpendicular to the weld surface, the joint W3 has a larger gradient of the boundary line, and the joint W2 has the largest degree of inclination. Obviously, there are hole defects at the bottom boundary of NZ and TMAZ on AS of both W1 and W3.

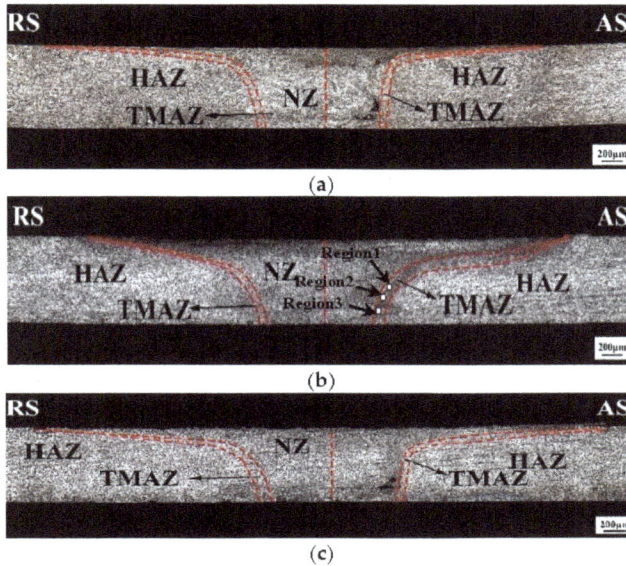

(a)

(b)

(c)

Figure 6. Macro-morphology of the weld: (**a**) W1; (**b**) W2; (**c**) W3.

There are some similarities between the three joints, as shown in Figure 6. The nugget zone (NZ) is darker than other zones, which is because the grains in NZ are finer than those in other zones. A clear demarcation exists between the nugget zone (NZ) and thermo-mechanically affected zone (TMAZ). TMAZ is located between NZ and the heat affected zone (HAZ). Nevertheless, there is no clear dividing line between HAZ and the base-metal (BM). At the same time, no obvious "S curve" is found in all joints. An "S curve" is usually caused by the surface oxidation film, which is not completely broken by the stirring during the traditional FSW process [27]. Three regions in AS were chosen to be scanned to detect the presence of an oxide film (Figure 6b). Oxide is obviously present and gathers in a small scale. Therefore, the aggregations are scattered (Figure 7). It can be seen that the stirring effect can be greatly improved in high rotation speed conditions. The oxide film rubbed against the shoulder and the pin of the FSW tool is completely crushed and stirred into the weld metal and cannot form a continuous distribution.

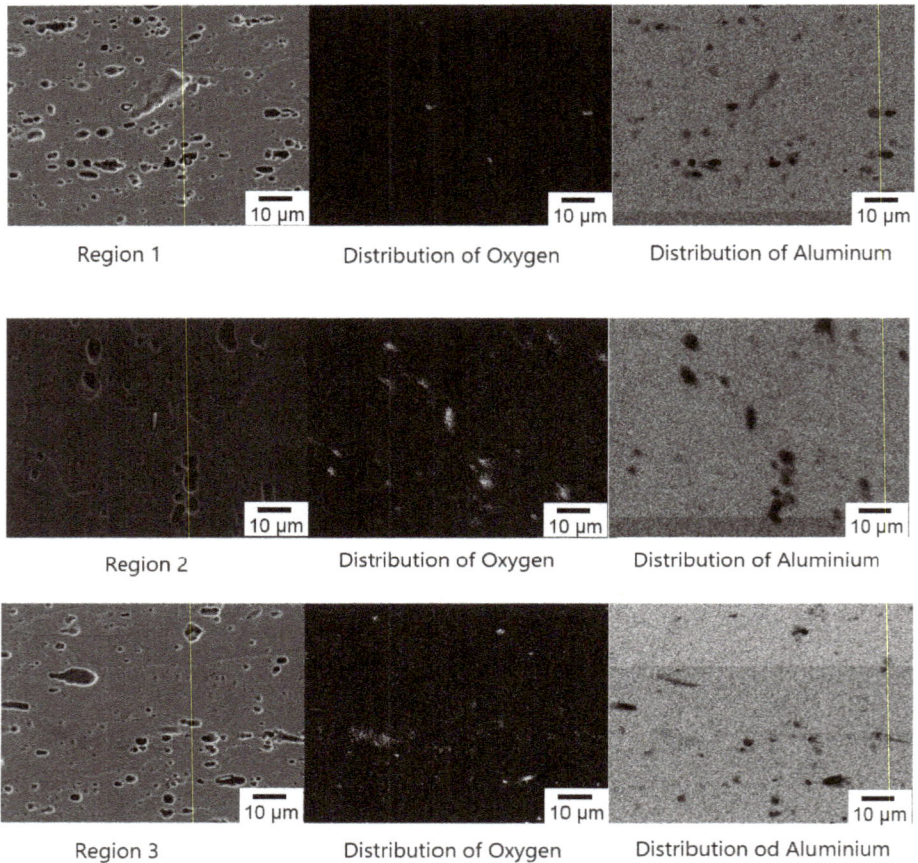

Figure 7. Micro-area scanning of oxides. The dark field image shows the oxide, and the bright field images highlight the aluminum.

The hardness curves of joints are shown in Figure 8. Three straight lines (L1, L2, L3) are selected. L2 is located in the weld center, the distance between the L1 and the upper surface is 0.2 mm, and the distance between the L3 and the bottom is 0.2 mm. The interval of two measurement points is 0.2 mm.

As displayed in Figure 8, the hardness curve of the weld cross-section is roughly "W". The hardness value in BM-HAZ-TMAZ-NZ declines first, and then gradually increases. This is mainly because HAZ is only subjected to thermal cycling and then the grains grow slightly bigger than that of BM. NZ is mainly composed of small equiaxed grains, so its hardness is higher than that of HAZ and TMAZ.

Figure 8d compares the average hardness of different welds. The hardness of all weld zones corresponding to W1 and W3 is lower than that of W2. In terms of the hardness distribution, the pin design of the quadrangular frustum pyramid results in more fully stirring and better joints.

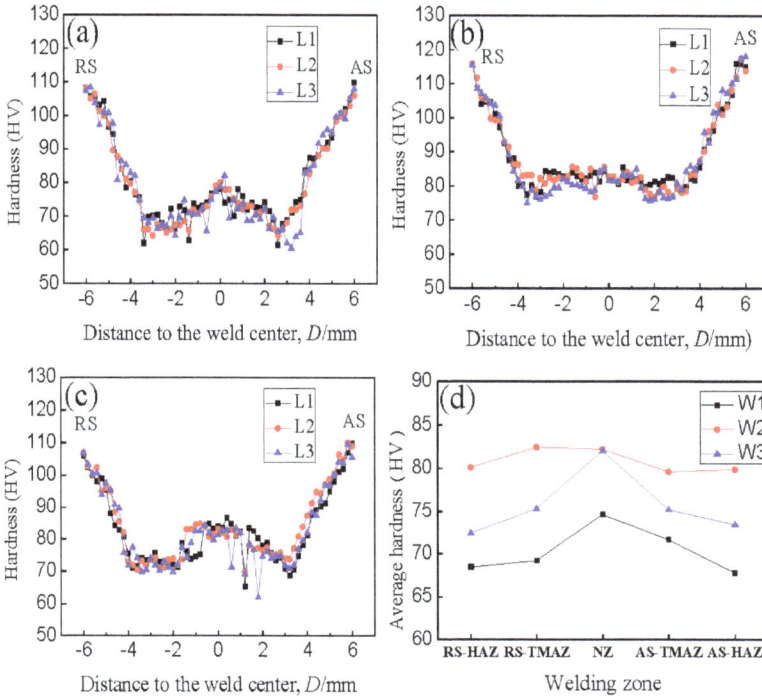

Figure 8. Hardness distribution characteristics of cross-section: (**a**) W1; (**b**) W2; (**c**) W3; (**d**) comparison of average hardness of each zone in W1, W2, and W3 welds.

3.4. Tensile Properties of the Joint

Tensile tests on the welded joints fabricated by S1, S2 and S3 FSW tools were carried out. The specimens were tested on Electro-mechanical Universal Testing Machines (Shandong Liangong Testing Machines Co., Ltd, Jinan, China), and the specimen dimensions are shown in Figure 9. The length of the specimen is 160 mm, the width of the specimen is 12 mm, the head width is 24 mm, and the transition radius is 30 mm. The test results are shown in Figures 10 and 11. The tensile strength of specimen W1, W2, and W3 is 160 MPa, 265 MPa, and 214 MPa, respectively, accounting for 52.6%, 87.2% and 70.4% of the base metal. Because the samples are not fully fixed to the testing machine, the stress stays nearly constant but the strain increases in Figure 10, which is inevitable.

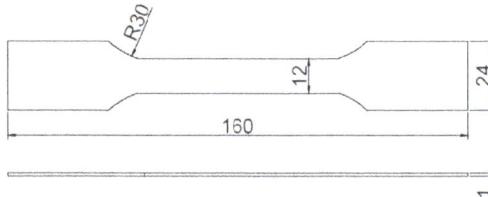

Figure 9. The dimensions of the tensile specimen (mm).

As is shown in Figure 11, the fracture locations and forms of joints are different. The fracture location of the specimen W1 is located at the junction of HAZ and NZ on the AS and the tensile strength is poor, and it is near the hole defects on W1. The fracture location of the specimen W2 is at the junction of HAZ and BM on the RS. The fracture location of the specimen W3 is located at the boundary between HAZ and TMAZ of AS close to the hole defects, with the tensile strength and elongation between W1 and W2. Obviously, the defects lead to a fracture with the tensile strength.

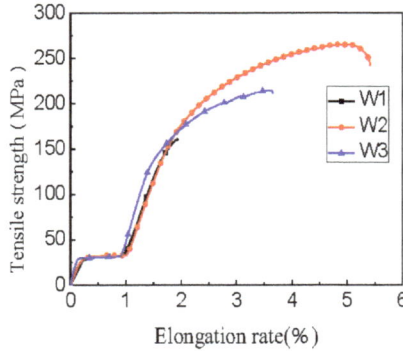

Figure 10. Tensile curves of specimens.

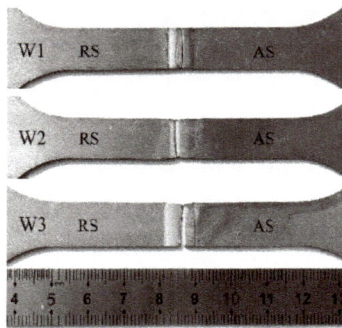

Figure 11. Fracture locations of joints.

4. Discussion

4.1. Microstructure of the Fracture

Because of the different shapes of the pins, the ability of the plastic metal to move along the vertical direction varies. The cross section of the weld W1, W2, and W3 is shown in Figure 12. The boundaries of the various zones of the AS are more obvious, but not in the RS.

For the specimen W1, there is an obvious hole defect at the bottom boundary of NZ and TMAZ, even if it has a smooth weld surface, as shown in Figure 12b. As the side of the prism is perpendicular to its bottom, the plastic metal is stirred only on the horizontal plane and fails to smoothly transition from NZ.

For a quadrangular frustum pyramid pin, the angle between its side and the bottom is an obtuse angle, the plastic metal flowing in the vertical and horizontal direction means that the plastic metal fully stirred, and the transition from TMAZ to NZ is smooth. As is shown in Figure 12c,d, the microstructures of TMAZ and NZ are compact, and almost no boundary line exists at the bottom of the weld.

As is shown in Figure 12f, hole defects are also generated in the specimen W3, even if the angle between the side of the pin and the bottom is also an obtuse angle. Due to the smooth outer surface, the conical pin is weaker on stirring in the horizontal direction in the welding process, and then it reduces the amount of plastic metals involved in the stirring. Naturally, an uneven transition between TMAZ and NZ occurs.

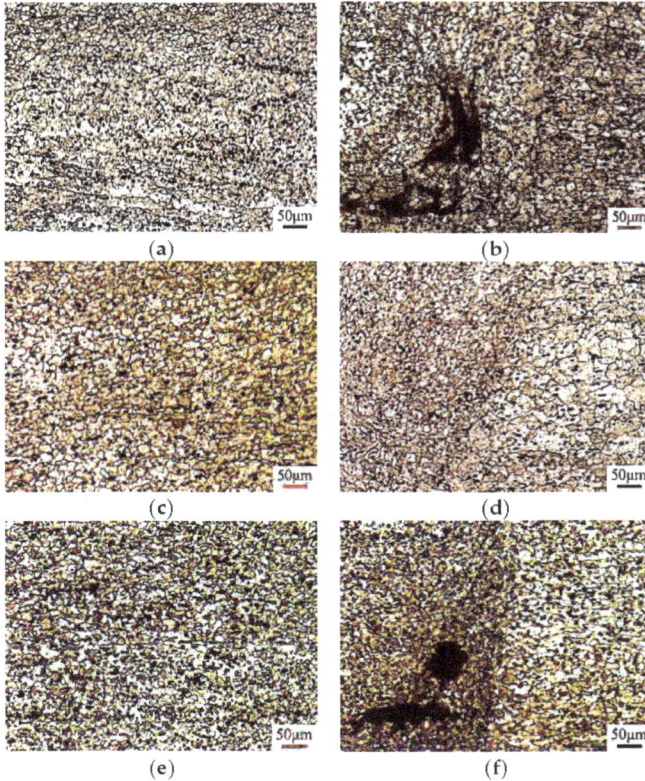

Figure 12. The microstructure of the weld fabricated by three different FSW tools: (**a**) the retreating side of W1 weld; (**b**) the advancing side of W1 weld; (**c**) the retreating side of W2 weld; (**d**) the advancing side of W2 weld; (**e**) the retreating side of W3 weld; (**f**) the advancing side of W3 weld.

A closer analysis revealed that the advancing side easily develops defects, which is mainly caused by an insufficient flow of plastic metals. As discussed above, we can note that the pin with four prisms cannot provide the driving force in the vertical direction and the weld is apt to defects. The conical pin possesses the ability to drive plastic metal in a vertical direction, but it is weaker on stirring in the horizontal direction in the welding process because of the smooth shape, and inevitably, the weld shows tunnel defects. The defects of these two shapes are both located at the junction of HAZ and NZ on the advancing side of the weld. On the contrary, the shape of the frustum pin generates a good weld and the fracture appears in the junction of HAZ and BM on the RS of the weld.

4.2. Grain Characteristics of HSFSW

In order to further reveal the grain characteristics by using a frustum pin, the Electron Backscattered Diffraction system (EBSD), was used to analyze the various zones of the joint. The grain orientation distribution, the grain size, and the grain deformation degree were analyzed at the same time.

The characteristics of BM are shown in Figure 13a. The grains are lath-like and their average diameter is 16.3 μm. There are also a large number of low angle boundaries ($2° < \theta < 15°$, θ is the grain boundary orientation angle), which is confirmed in Figure 14a. As shown in Figure 13b, HAZ is still dominated by small angle grain boundaries and the average grain diameter is 16.9 µm because of the heat cycle during the welding process.

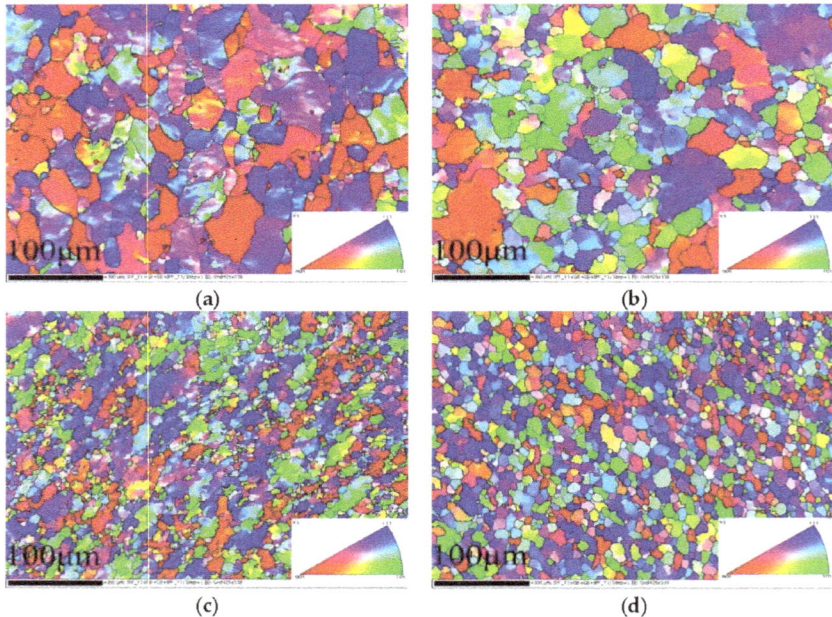

Figure 13. OIM (Orientation Imaging Microscopy) photographs of welded joints (W2): (**a**) BM; (**b**) HAZ; (**c**) TMAZ; (**d**) NZ.

The NZ is composed of equiaxed grains (the average grain diameter is 9.3 µm) with high angle boundaries ($\theta > 15°$) in Figure 13d. The transformation from low angle grain boundaries continuously increases the number of high angle grain boundaries and finally the grains are significantly refined. The composition of the TMAZ is similar to that of the NZ, but there are more deformed grains in the TMAZ, and the grains are irregular (Figure 13c).

For the traditional FSW, the average grain size of HAZ is 18.2–18.9 µm, and the average grain size of NZ is 9.1–9.7 µm [28]. That is to say, the high speed does not produce too much heat and the grain size of the HSFSW joint is similar to that of traditional FSW.

4.3. Grain Orientation Distribution Map

The orientation difference between the adjacent grains of the deformed structure also affects the deformation and fracture behavior of the weld seam. Owing to a large number of deformed structures inside the aluminum alloy sheet before welding (Figure 14a), the distortion energy is very low, and dynamic recrystallization does not occur here. The grain orientation distribution of HAZ is shown in Figure 14b. Similar to that of BM (57.5%), a large number of low angle boundaries still exist, but the proportion slightly decreases to 50%. This is because HAZ is mainly affected by thermal cycling. The grain structure only slightly grows along the deformation direction and the orientation difference almost does not change.

Figure 14c shows the grain orientation distribution of TMAZ. The proportion of low angle boundaries is 23.3%, which shows that the proportion of high angle boundaries increases and the grain structure has undergone significant changes. Large plastic deformation exits near the nugget zone, subjected to both thermal cycling and shearing stress, so the internal energy and atomic activity increase. With the slip and accumulation of dislocations, subgrain boundaries with extremely low misorientation were finally produced. Subgrain boundaries firstly changed into low angle boundaries, and then gradually transformed into high angle boundaries owing to continuous dynamic recovery and recrystallization. Therefore, the equiaxed grains surrounded by high angle boundaries have no sub-structure. However, the grains away from NZ are mainly affected by the thermal cycle and still have low angle boundaries.

Compared with BM, the proportion of low angle boundaries in NZ is significantly reduced to 20.1% (Figure 14d). According to the existing research on the traditional FSW joints of the 1 mm 6061-T6 aluminum alloy, the proportion of low angle boundaries in HAZ accounted for 70.5%, in TMAZ accounted for 60.3%, and in NZ accounted for 55.46% [12,29]. Obviously, the proportion of low angle boundaries in all zones in our research is lower than that of the traditional FSW, which indicates that the number of high angle boundaries can effectively hinder the crack expansion and greatly improve the connection strength of the weld. By high rotation speed, the grains in the NZ are simultaneously subjected to the squeezing and stronger shearing force of the tool, the dislocation density increases continuously, and the orientation deviation of low angle boundaries increases. Once recrystallization occurs, new equiaxed grains emerge and the low angle boundaries are soon changed into high angle boundaries. However, the crystal nucleus in NZ was mechanically broken without an increase, and was then transformed into small equiaxed grains, which have much a smaller diameter than that of BM.

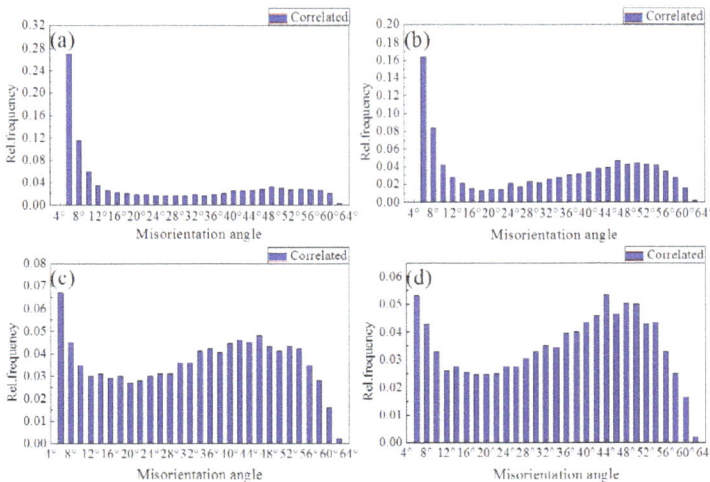

Figure 14. Orientation distribution of 6061-T6 aluminum alloy welded joints (W2): (**a**) BM; (**b**) HAZ; (**c**) TMAZ; (**d**) NZ. Rel. (relative) frequency is the probability of occurrence of the grain boundary with various orientation angles.

5. Conclusion

Three kinds of pins are selected to perform high rotation speed FSW on a 1 mm 6061-T6 aluminum alloy sheet. When the shape of the pin is a quadrangular frustum pyramid, the rotation speed is 11,000 rpm and the travel speed is 300 mm/min, and soundable joints are obtained. Due to the stirring effect of high rotation speed, the proportion of low angle boundaries in all zones is lower than that of the traditional FSW, while the average grain size is similar to traditional FSW. The proportion of low angle boundaries in HAZ, TMAZ, and NZ is 50%, 23.3%, and 20.1%, respectively. The tensile strength of specimen W2 is 265 MPa, which accounts for 87.2% of the base metal. The pattern of the weld cross section is "flat T". HAZ and TMAZ are narrow and no obvious "S curve" is found in the weld, which is different from the traditional FSW.

Author Contributions: Data curation formal analysis, writing-original draft, and Writing—review & editing, Y.Z.; methodology and conceptualization, J.W.; funding acquisition, S.C.; project administration, P.W.; resources and investigation, J.X.

Funding: This research was funded by the Qing Lan Project, National Post Doctoral Fund and the National Natural Science Foundation of China (51675248) and the Natural Science Fund of the Jiangsu Higher Education Institutions of China (17KJA460006).

Acknowledgments: The authors would like to thank the support of the laboratories and the help of Wu Yunkai for writing the draft.

Conflicts of Interest: The authors declare no conflict of interest.

References

1. Palanivel, R.; Mathews, P.K.; Murugan, N.; Dinaharan, I. Effect of tool rotational speed and pin profile on microstructure and tensile strength of dissimilar friction stir welded AA5083-H111 and AA6351-T6 aluminum alloys. *Mater. Des.* **2012**, *40*, 7–16. [CrossRef]
2. Sasabe, S.; Eguchi, N.; Ema, M.; Matsumoto, T. Laser welding characteristics of aluminium alloys for automotive applications. *Weld. Int.* **2003**, *17*, 870–878. [CrossRef]
3. Haraga, K.; Kanesaka, T.; Mabuchi, A. Strength properties of aluminum/aluminum and aluminum/steel joints for light weighting of automotive body. *Weld. World* **2000**, *44*, 23–27.
4. Liang, H.M.; Yan, K.; Wang, Q.Z.; Zhao, Y.; Liu, C.; Zhang, H. Improvement in joint strength of spray-deposited Al-Zn-Mg-Cu alloy in underwater friction stir welding by altered temperature of cooling water. *J. Mater. Eng. Perform.* **2016**, *25*, 1–8. [CrossRef]
5. Çam, G.; Mistikoglu, S. Recent developments in friction stir welding of Al-alloys. *J. Mater. Eng. Perform.* **2014**, *23*, 1936–1953. [CrossRef]
6. Song, K.H.; Tsumura, T.; Nakata, K. Development of microstructure and mechanical properties in Laser-FSW hybrid welded Inconel 600. *Mater. Trans.* **2009**, *50*, 1832–1837. [CrossRef]
7. Song, S.W.; Kim, B.C.; Yoon, T.J.; Kim, N.K.; Kim, I.B.; Kang, C.Y. Effect of welding parameters on weld formation and mechanical properties in dissimilar Al alloy joints by FSW. *Mater. Trans.* **2010**, *51*, 1319–1325. [CrossRef]
8. Rhodes, C.G.; Mahoney, M.W.; Bingel, W.H.; Spurling, R.A.; Bampton, C.C. Effects of friction stir welding on microstructure of 7075 aluminum. *Scr. Mater.* **1997**, *36*, 69–75. [CrossRef]
9. Sutton, M.A.; Yang, B.; Reynolds, A.P.; Taylor, R. Microstructural studies of friction stir welds in 2024-T3 aluminum. *Mater. Sci. Eng. A* **2002**, *323*, 160–166. [CrossRef]
10. Lu, S.; Jia, X.D.; Zhang, C.Y.; Fu, L.; Dong, X.Y. Temperature field and microstructure of magnesium alloy fabricated by FSW. *J. Mater. Eng. Perform.* **2009**, *1*, 9–13.
11. Scialpi, A.; Giorgi, M.D.; Filippis, L.A.C.D.; Nobile, R.; Panella, F.W. Mechanical analysis of ultra-thin friction stir welding joined sheets with dissimilar and similar materials. *Mater. Des.* **2008**, *29*, 928–936. [CrossRef]
12. Tong, J.H.; Li, L.; Deng, D.; Wan, F.R. Friction stir welding of 6061-T6 aluminum alloy thin sheets. *J. Univ. Sci. Technol. Beijing* **2008**, *30*, 1011–1017.
13. Serio, L.M.; Palumbo, D.; De Filippis, L.A.C.; Galietti, U.; Ludovico, A.D. Effect of Friction Stir Process Parameters on the Mechanical and Thermal Behavior of 5754-H111 Aluminum Plates. *Materials* **2016**, *9*, 122. [CrossRef] [PubMed]
14. De Filippis, L.A.C.; Serio, L.M.; Palumbo, D.; De Finis, R.; Galietti, U. Optimization and characterization of the Friction Stir Welded Sheets of AA 5754-H111: Monitoring of the quality of joints with thermographic techniques. *Materials* **2017**, *10*, 1165. [CrossRef] [PubMed]
15. Zhao, H.H.; Feng, X.S.; Xiong, Y.Y.; Li, H.J.; Dong, F.B.; Hu, L.; Guo, L.J. Study on the temperature distribution of 6061 aluminium alloy micro friction stir welding featured high speed without inclination. *Electr. Weld. Mach.* **2014**, *44*, 71–77.
16. Qin, G.L.; Zhang, K.; Zhang, W.B.; Wu, C.S. Effect of friction stir welding heat input on weld appearance and mechanical properties of 6013-T4 Al alloy joint. *Trans. China Weld. Inst.* **2010**, *31*, 5–8.

17. Tian, Z.J.; Su, Z.Q.; Gao, Y.J.; Xu, S.J.; S, S.X. Research on FSW and VPPA intercross welding of 2219 aluminum alloy. *Weld. Technol.* **2013**, *42*, 24–26.
18. Chen, S.J.; Zhou, Y.; Xue, J.R.; Ni, R.Y.; Guo, Y.; Dong, J.H. High rotation speed friction stir welding for 2014 aluminum alloy thin sheets. *J. Mater. Eng. Perform.* **2017**, *26*, 1–9. [CrossRef]
19. Chen, Y.C.; Liu, H.J.; Feng, J.C. Friction stir welding characteristics of different heat-treated-state 2219 aluminum alloy plates. *Mater. Sci. Eng. A* **2006**, *420*, 21–25. [CrossRef]
20. Elangovan, K.; Balasubramanian, V. Influences of pin profile and rotational speed of the tool on the formation of friction stir processing zone in AA2219 aluminiumalloy. *Mater. Sci. Eng. A* **2007**, *459*, 7–18. [CrossRef]
21. Liu, H.J.; Chen, Y.C.; Feng, J.C. Effect of zigzag line on the mechanical properties of friction stir welded joints of an Al–Cu alloy. *Scr. Mater.* **2006**, *55*, 231–234. [CrossRef]
22. Zhang, Z.; Liu, H.J. Effect of pin shapes on material deformation and temperature field in friction stir welding. *Trans. China Weld. Inst.* **2011**, *32*, 5–8.
23. Galvão, I.; Leal, R.M.; Rodrigues, D.M.; Loureiro, A. Influence of tool shoulder geometry on properties of friction stir welds in thin copper sheets. *J. Mater. Process. Technol.* **2013**, *213*, 129–135. [CrossRef]
24. Faraji, G.; Asadi, P. Characterization of AZ91/alumina nanocomposite produced by FSP. *Mater. Sci. Eng. A* **2011**, *528*, 2431–2440. [CrossRef]
25. Azizieh, M.; Kokabi, A.H.; Abachi, P. Effect of rotational speed and probe profile on microstructure and hardness of AZ31/Al$_2$O$_3$, nanocomposites fabricated by friction stir processing. *Mater. Des.* **2011**, *32*, 2034–2041. [CrossRef]
26. Chen, Z.; Zhou, Y.L.; Tian, B.; Zhang, T.; Liu, Y.J. Research on FSW of 6061 aluminum alloy. *Electr. Weld. Mach.* **2011**, *41*, 95–98.
27. Xie, T.F.; Xing, L.; Ke, L.M.; Luan, G.H.; Dong, C.L. Influence of Pin geometry on formation of lazy S in Friction Stir Welding. *Hot Work. Technol.* **2008**, *37*, 64–66.
28. Jia, Y.; Wang, K.H.; Yang, L.; Wang, X.J. Analysis on Microstructure of 6061 Al Alloy with Friction Stir Welding. *Hot Work. Technol.* **2015**, *44*, 180–182.
29. Wang, B.; Lei, B.B.; Zhu, J.X.; Feng, Q.; Wang, L.; Wu, D. EBSD study on microstructure and texture of friction stir welded AA5052-O and AA6061-T6 dissimilar joint. *Mater. Des.* **2015**, *87*, 593–599. [CrossRef]

Article

Performance of Plunge Depth Control Methods During Friction Stir Welding

Jinyoung Yoon [1,2], Cheolhee Kim [1,3,*] and Sehun Rhee [2]

[1] Joining Research Group, Korea Institute of Industrial Technology, Incheon 21999, Korea; 0521jin@kitech.re.kr
[2] School of Mechanical Engineering, Hanyang University, Seoul 04763, Korea; srhee@hanyang.ac.kr
[3] Department of Mechanical and Materials Engineering, Portland State University, Portland, OR 97201, USA
* Correspondence: chkim@kitech.re.kr; Tel.: +82-32-850-0222

Received: 13 February 2019; Accepted: 27 February 2019; Published: 2 March 2019

Abstract: Friction stir welding is a preferred solid state welding process for Al/Fe joints, and in friction stir lap welding, the plunge depth is the most critical parameter for joint strength. We compared three plunge depth control methods, namely conventional position control, offset position control, and deflection compensation control in the friction stir lap welding of 3 mm-thick Al 5083-O alloy over 1.2 mm-thick DP 590 steel. The desired plunge depth was 0.2 mm into the steel sheet. However, the pin did not reach the steel surface under conventional position control due to deflection of the vertical axis of the welding system. In offset position control, an additional offset of 0.35 mm could achieve the desired plunge depth with considerable accuracy. Nevertheless, a gradual increase of the plunge depth along the longitudinal direction was unavoidable, due to an in-situ decrease of the material yield strengths. In deflection compensation control, the deflection is estimated by the coaxially measured plunging force and the force-deflection relationship, and then corrected by feedback control. Thus, the plunge depth is stabilized along the longitudinal direction and is precisely controlled with a 3.3-μm standard deviation of error during the tool traverse phase. There is also a consistent bias of 32 μm caused by the resolution of the measuring system, and it can be easily calibrated in the feedback control system.

Keywords: Al/Fe dissimilar joining; friction stir welding; plunge depth control; offset position control; deflection compensation control

1. Introduction

In the automotive industry, there is rapidly increasing use of high-strength steels and aluminum to reduce the weight of vehicles. To improve the performance and price competitiveness of automobiles, the development of a multi-materials car body, which adopts various materials simultaneously into the parts, is an important issue. Thus, there is growing research interest in the joining technology of different materials [1,2]. Steel together with Al alloy is considered the most important dissimilar material combination, for which various approaches have been investigated [3]. During the fusion welding of steel and Al alloy, a low heat input process is preferred because the joining strength is reduced by the formation of an intermetallic compound (IMC). Galvanic corrosion is another critical issue for the Al/Fe combination as it hinders the durability of the joint [4–6]. Currently, the preferred industrial methods are adhesive bonding and mechanical joining techniques, such as riveting and clinching, because of no IMC formation and high galvanic corrosion [3,7]. However, there is a continuous demand for more economical welding techniques. Solid state welding can drastically reduce IMC formation and ensure high bonding strength compared to fusion welding. The Honda Motor Company successfully applied friction stir welding (FSW) for dissimilar metals of Al/Fe in the commercial mass production of Accord 2013 model cars [8]. They applied robotic FSW on Al/Fe overlap joints to weld the latter by forming a thin IMC layer of Fe_4Al_{13}. In this case, the pin on the FSW tool plunged through the upper Al part and slightly penetrated the top of the lower steel part.

Various FSW techniques for Al/Fe joints and the resultant joint properties were extensively reviewed by Hussein et al. [9]. Those authors classified the techniques into three types: Diffusion, plunging, and annealing; and described the characteristics of various processes. The entire FSW sequence can also be divided into three phases in time: The plunge and dwell, tool traverse, and retract phases [10]. In terms of the joining strength, the position of the pin during the plunge and traverse phases, called the plunge depth, is the most important parameter. Kimapong et al. first implemented FSW on Al/Fe butt joints, and the highest joining strength (about 86% of that of the Al base material) was achieved when the pin was mostly on the Al side with 0.2 mm inserted into the steel side [11]. In the friction stir lap welding (FSLW) of Al/Fe joints, Al is usually placed on top of the steel, and the pin penetrates the lower steel plate by 0.1 to 0.2 mm [12–14]. The change of the plunge depth influences the joining strength by either changing the shape of the "hook" formed at the interface [15] or inducing excessive IMC growth by heat generation [16]. Excessive plunge depths can also cause tool wear and reduce the tool life. Therefore, it is very important to maintain a small penetration depth into the lower steel sheet in the FSLW of the Al/Fe overlap joint.

The FSW system can be implemented by conventional machine tools, dedicated FSW machines, or industrial robots [17]. In all these systems, the tool position is basically controlled by pure position control. In the FSLW of the Al/Fe joint, when the pin tip penetrates the steel, there is a higher axial load and subsequent deflection of the system. This system deflection is not compensated for by the position control, and so both the plunge depth and joining strength are reduced. Smith [18] and Cook et al. [19] reported that deflection in the FSLW system could be reduced and the welding quality enhanced by the constant plunging force control. In their subsequent research, Cook and coworkers [20] evaluated the plunge depth, traverse speed, and rotation speed as control parameters to maintain a constant plunging force. Although the plunging force control can considerably compensate for system deflection compared to conventional position control, it has two drawbacks. First, because the plunging force is affected by not only the plunge depth, but also other process parameters (such as the traverse speed and rotation speed), the plunging force required to reach a certain plunge depth varies with the process parameters [17,19,20]. Second, even if a constant plunging force can be maintained, the actual plunge depth may be inappropriate when the yield strength of the base materials changes with temperature. Thus, Smith and coworkers [21] implemented simultaneous temperature control with the plunging force control in order to improve the joint quality.

The offset position control is carried out by adding a certain offset to the reference position. Being simpler than the constant plunging force control, it is often applied to compensate for the system deflection. However, characteristics of this control process have not been reported so far. Very recently, our group reported a force-deflection model to compensate for the position error in friction stir spot welding (hereafter called the deflection compensation control) [22]. The axial load, i.e., the plunging force was coaxially measured using a load cell, and the deflection estimated by the force-deflection model was compensated for. Importantly, the relationship between the axial load and the system deflection depends only on the stiffness of the welding system rather than the base materials or process parameters. By using the proposed model, the plunge depth could be controlled with an error of less than 50 μm for various process parameter sets. In this paper, we further compare the deflection compensation control to the conventional position control and the offset position control methods in terms of effectively controlling the plunge depth and joint properties.

2. Experimental Setup

The base materials were Al 5083-O alloy with a thickness of 3.0 mm and dual phase (DP) 590 steel with a thickness of 1.2 mm. Their chemical compositions are given in Table 1. The welding tool was made of tungsten carbide with 12% Co; and the pin length, pin diameter, and shoulder diameter were 2.7, 3.0, and 12 mm, respectively. Details of the tool shape were given in the previous paper [22].

As shown in Figure 1, the Al alloy sheet is overlaid on the steel sheet for the FSLW, and the welding tool is tilted by 3° against the welding direction. The desired plunge depth is set at 3.2 mm, where the pin of the welding tool penetrates the bottom sheet by 0.2 mm. An insufficient plunging leads to a reduced interface area, and excessive plunging causes excess IMC formation [14,16,23,24].

Table 1. Chemical composition of base materials (wt%).

Al 5083-O								
Si	**Fe**	**Cu**	**Mn**	**Mg**	**Cr**	**Zn**	**Ti**	**Al**
0.14	0.26	0.04	0.69	4.54	0.11	0.01	0.02	Bal.

DP 590					
C	**Si**	**Mn**	**P**	**S**	**Fe**
0.078	0.362	1.809	0.0172	0.0014	Bal.

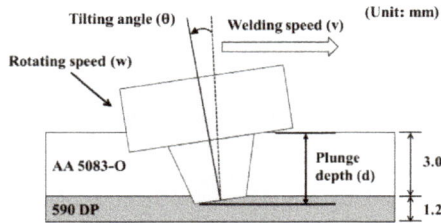

Figure 1. Specimen and welding tool configuration.

The conventional position control and offset position control were implemented by a computer numerical control (CNC) machine controller in the 3-axis cartesian FSW system. The deflection compensation control was implemented using co-axial load measurement and feedback control (Figure 2). In the feedback control, the axial deflection of the FSW system is estimated and compensated for by using a linear load-deflection relationship. Details of the measurement system, control system, and the control algorithm can be found in the previous paper [22].

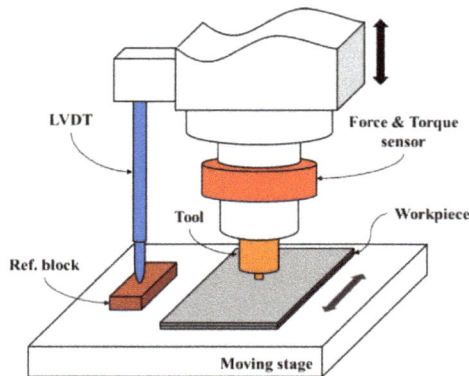

Figure 2. Schematic diagram of the deflection compensation control system.

In the position control and the deflection compensation control, the tool rotation speed and welding speed are selected as process parameters while the plunging speed and dwell time are fixed at 20 mm/min and 7 s, respectively. The tool rotation speed has 3 levels (500, 700, and 900 rpm) and the welding speed also has 3 levels (100, 150, and 200 mm/min) in the experiments. The actual plunge depth was recorded by a linear variable differential transformer (LVDT) sensor, and the force and torque were recorded by a coaxial sensor. For the offset position control, the tool rotation speed and welding speed were fixed at 700 rpm and 150 mm/min, respectively, while only the offset value to compensate for the axial deflection was varied from 0.20 to 0.55 mm.

Five tensile shear specimens were prepared for each condition according to ISO 6892, with a gage length of 60 mm, a gage width of 12.5 mm, and an overlap length of 50 mm. The load upon fracture was measured under a test speed of 5 mm/min.

3. Results and Discussion

3.1. Conventional Position Control

When using the conventional position control, none of the nine parameters sets could allow the pin to penetrate the lower steel sheet at all, while a penetration depth of 0.2 mm is desired (Figure 3). This insufficient penetration is caused by deflection of the system, which is intrinsically not compensated for in this case.

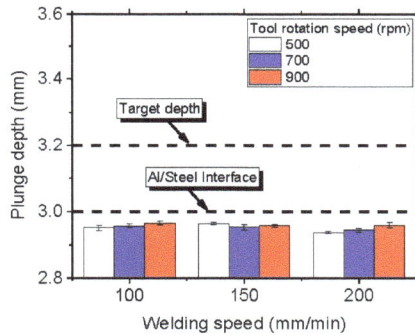

Figure 3. Measured plunge depth under the conventional position control.

Nevertheless, the plunge depth slightly decreases with the advance per revolution (APR), which is defined by the ratio of the welding speed to the tool rotation speed (Figure 4) [10]. The concept of APR is similar to the reciprocal of the heat input per unit length in convention fusion welding. The lower the APR, the higher the temperature and the lower the yield strength of the base materials. In these experiments, the plunge depth varies within only a small range of 55 μm, because the entire welding pin remains within the upper Al sheet.

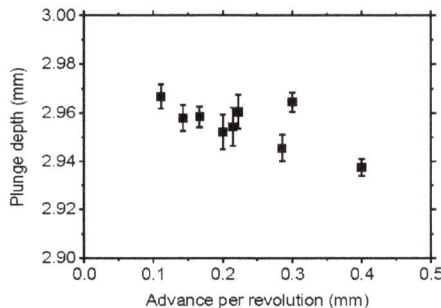

Figure 4. Plunge depth according to advance per revolution under the conventional position control.

3.2. Offset Position Control

When an additional offset of 0.20–0.55 mm is assigned in the position control to compensate for the vertical deflection of the system, the measured plunge depth linearly increases from 3.11 to 3.38 mm (Figure 5), while the desired plunge depth is 3.2 mm. The slope and intercept of the linear fitting line are 0.784 and 2.93 mm, respectively. Note that the slope is not in unity and can vary with the process parameters. For this reason, the adequate offset for a given plunge depth is hard to estimate, and it used to be determined by experimental trial and error.

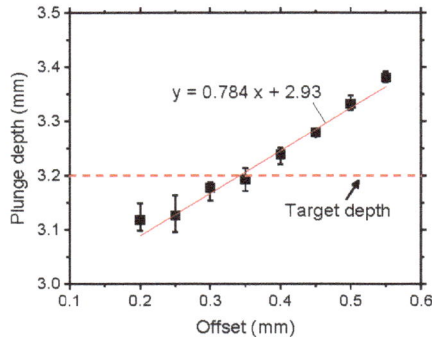

Figure 5. Plunge depth according to the additional offset under the offset position control (welding speed: 150 mm/min, rotation speed: 700 rpm).

When the plunge depth is set to 3.55 mm in the position control system, i.e., an offset of 0.35 mm, the actual plunge depth is nearest to the desired value of 3.2 mm (Figure 5). However, even for this case, the measured plunge depth fluctuates with time, and the variation range is 42 μm (Figure 6) during the tool traverse phase. At the same time, the axial force continuously decreases from 7.4 to 6.1 kN assuming the linear fit. Because the temperature of the specimen increases with time, a lower force and a higher plunge depth were observed. The deflection can be calculated from the measured force by using the force-deflection model developed in the previous study [22]. When calculated with the given change of the axial force, the difference in the system deflections before and after the tool traverse phase is estimated as 45 μm, which agrees very well with the measured difference in the plunge depth (42 μm).

Tensile shear test was conducted for specimens taken at every 20 mm along the weld bead (Figure 7). The maximum fracture load is 4.2 kN, and the range of the load is 0.9 kN. The gradual decrease of the fracture load can originate from the increasing plunge depth and temperature, which promote growth of IMC during the joining of Al/Fe metals [16,25]. The difference in the fracture load at different positions is more than 20%, which can cause an overdesign of welds and a decease in the productivity and quality of the process.

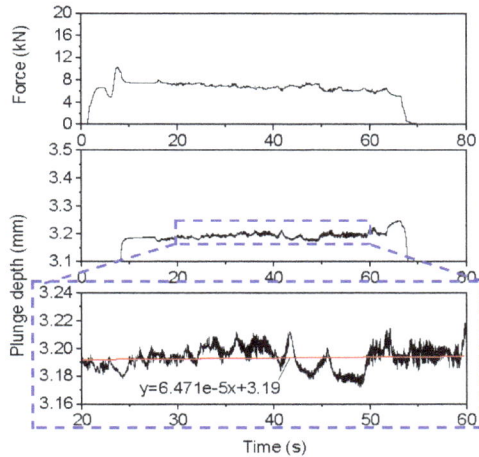

Figure 6. Measured profiles of the plunge depth and axial force under the offset position control (welding speed: 150 mm/min, rotation speed: 700 rpm, offset: 0.35 mm).

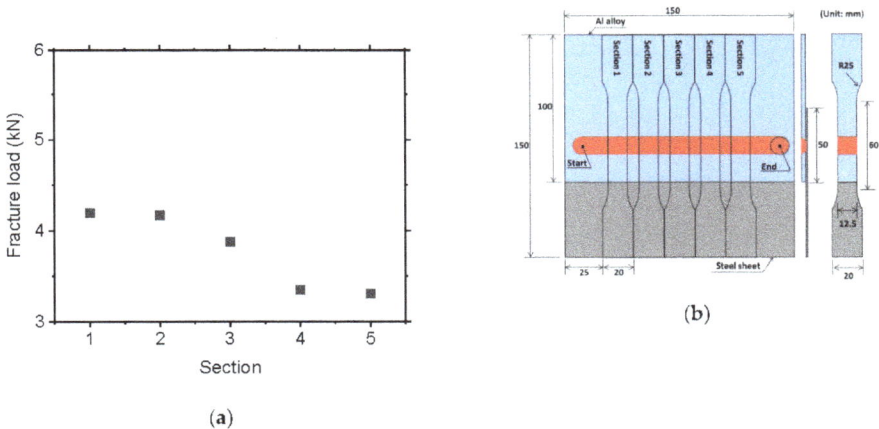

(a)

(b)

Figure 7. Tensile shear test results according to the location of specimens under the offset position control; (a) fracture load; (b) specimen preparation (welding speed: 150 mm/min, rotation speed: 700 rpm, offset: 0.35 mm).

3.3. Deflection Compensation Control

When the deflection compensation control is applied, both the plunge depth and axial force are stabilized during the entire tool traverse phase (Figure 8). The average value and range of the plunge depth are 3.23 mm and 14 μm, respectively, while the desired depth is 3.2 mm. A bias of around 30 μm was observed and attributed to the resolution of the load measuring system, similar to the previous study [22]. The plunge depth is well controlled under various welding speeds and tool rotation speeds (Figure 9); for the nine parameter sets used in this experiment, the average standard deviation is only 3.3 μm with the error defined as the difference between the measured and desired plunge depths. The plunge depth is biased in a positive direction as shown in Figure 8, and the average bias for all cases is 32 μm. Because this bias is nearly constant regardless of the process parameters, it can be easily removed using proper calibration in the feedback system.

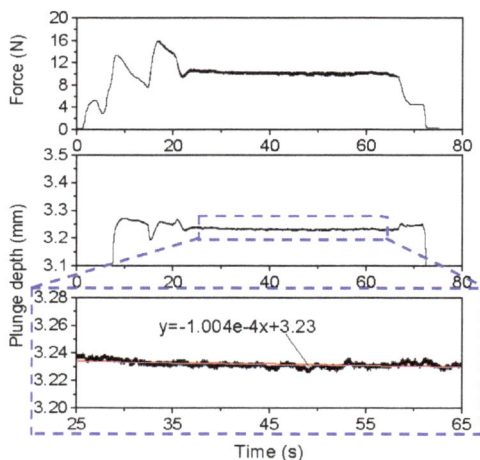

Figure 8. Measured profiles of the plunge depth and axial force under the deflection compensation control (welding speed: 150 mm/min, rotation speed: 700 rpm, offset: 0.35 mm).

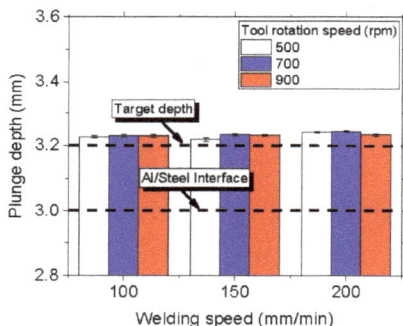

Figure 9. Measured plunge depth under the deflection compensation control.

The plunge depth under deflection compensation control is plotted with respect to the APR in Figure 10. Unlike the conventional position control in Figure 3, the plunge depth is unaffected by the APR and is maintained for all welding conditions. In the tensile shear test, the averaged fracture load is 6.3 kN with a range of 0.5 kN (Figure 11). In comparison, the fracture load in the offset position control decreases from 4.2 kN to 3.3 kN (by 0.9 kN) along the sectioning position (Figure 7). Therefore, the average fracture load increases by 68% and its range decreases by more than 44% by using the deflection compensation control.

Both the axial force and the torque increase linearly with the APR for almost the same plunge depth (Figure 12). A lower APR means more heat input into the base materials to increase their temperature. Consequently, the yield strength of the base materials decreases, and a lower axial force is required to achieve a fixed plunge depth. On the other hand, if a constant axial force or torque is applied, the plunge depth decreases with the APR. Therefore, the proper reference force or torque should be selected for a given APR, just like the selection of a proper offset for the process parameters in the offset position control.

Even more, the in-situ increase of temperature during the welding process can change the plunge depth along the longitudinal direction when the constant force or torque control method is applied. For example, as shown in Figure 8, the axial force decreases from 10.6 to 9.8 kN during the tool traverse phase under almost the same plunge depth (with a range of only 14 μm). This means that a constant force or torque control method cannot guarantee longitudinal consistency in the plunge depth, because the yield strength of the base materials decreases along the weld during the tool traverse phase owing to continuous heating.

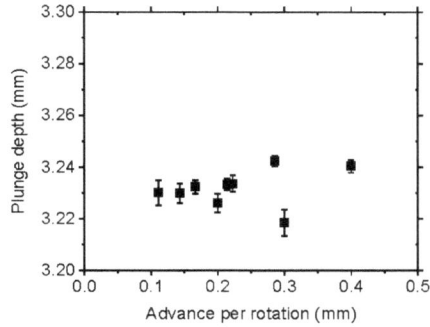

Figure 10. Plunge depth according to advance per revolution under the deflection compensation control.

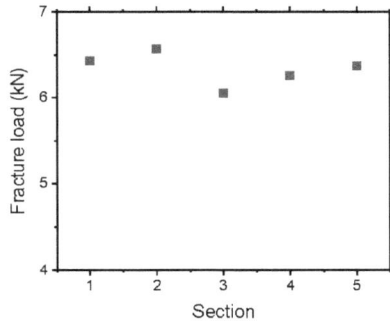

Figure 11. Tensile shear test results under the deflection compensation control according to the location of specimens (welding speed: 150 mm/min, rotation speed: 700 rpm, offset: 0.35 mm).

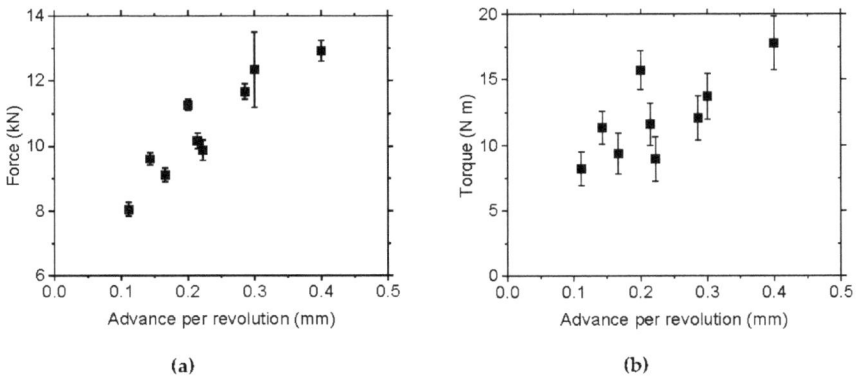

Figure 12. Axial force and torque according to advance per revolution under the deflection compensation control.

4. Conclusions

In this study, three plunge depth control methods (conventional position control, offset position control, and deflection compensation control) were examined in the friction stir lap welding of dissimilar Al alloy and steel sheets. A 3 mm-thick Al 5083-O alloy was used on the top and 1.2 mm-thick DP 590 steel was on the bottom. The performance of each control method in maintaining the desired pin plunge depth (3.2 mm penetration in the steel) and the joining strength was compared. The following conclusions were derived.

(1) When using the conventional position control, the actual plunge depth was below 3.0 mm. The pin could not reach the upper surface of the steel sheet due to system deflection, and proper welds were not established.

(2) In the offset position control experiments, the desired plunge depth was most accurately achieved by applying an addition offset of 0.35 mm, when the welding speed and tool rotation speeds were at 150 mm/min and 700 rpm, respectively. However, the plunge depth continuously increased by 42 μm during the tool traverse phase, and the corresponding fracture load in the tensile test decreased from 4.2 to 3.3 kN due to increased heat input during the welding.

(3) When the deflection compensation control was applied, precise control of the plunge depth was accomplished with a 3.3-μm standard deviation of error during the tool traverse phase. A bias of 32 μm into the DP steel was observed due to the resolution of the load sensor. This bias did not vary with the process parameters and could be easily removed to improve the control accuracy.

(4) Temperature changes in the base materials causes in-situ variation of the system deflection during the tool traverse phase. The deflection compensation control method can adequately compensate for this variation, which is not compensated for by the offset position control, force control, or torque control.

Author Contributions: Investigation, J.Y. and C.K.; Methodology, J.Y. and C.K.; Supervision, C.K. and S.R.; Writing—original draft, J.Y.; Writing—review & editing, C.K.

Funding: This research was supported by the Ministry of Trade, Industry and Energy, Republic of Korea.

Acknowledgments: The authors would like to thank Dr. Young-Pyo Kim and staffs of Hwacheon Machinary for their kind assistance with tailoring the friction stir welding system.

Conflicts of Interest: The authors declare no conflict of interest

References

1.   Meschut, G.; Janzen, V.; Olfermann, T. Innovative and highly productive joining technologies for multi-material lightweight car body structures. *J. Mater. Eng. Perform.* **2014**, *23*, 1515–1523. [CrossRef]

2.   Martinsen, K.; Hu, S.J.; Carlson, B.E. Joining of dissimilar materials. *Cirp Ann.* **2015**, *64*, 679–699. [CrossRef]

3.   Sakayama, T.; Naito, Y.; Miyazakki, Y.; Nose, T.; Murayma, G.; Saita, K.; Oikawa, H. Dissimilar metal joining technologies for steel sheet and aluminum alloy sheet in auto body. *Nippon Steel Tech. Rep.* **2013**, *103*, 91–98.

4.   Kang, M.J.; Kim, C.H. Cold-Metal-Transfer Arc Joining of Al 6K32 Alloy to Steel Sheets. In *Defect and Diffusion Forum*; Trans Tech Publications: Zurich, Switzerland, 2013; pp. 247–251.

5.   Kang, M.; Kim, C.; Kim, J.; Kim, D.; Kim, J.H. Corrosion assessment of Al/Fe dissimilar metal joint. *J. Weld. Join.* **2014**, *32*, 55–62. [CrossRef]

6.   Kang, M.; Kim, C. Joining Al 5052 alloy to aluminized steel sheet using cold metal transfer process. *Mater. Des.* **2015**, *81*, 95–103. [CrossRef]

7.   Groche, P.; Wohletz, S.; Brenneis, M.; Pabst, C.; Resch, F. Joining by forming—A review on joint mechanisms, applications and future trends. *J. Mater. Process. Technol.* **2014**, *214*, 1972–1994. [CrossRef]

8.   Kusuda, Y. Honda develops robotized FSW technology to weld steel and aluminum and applied it to a mass-production vehicle. *Ind. Robot Int. J.* **2013**, *40*, 208–212. [CrossRef]

9.   Hussein, S.A.; Tahir, A.S.M.; Hadzley, A. Characteristics of aluminum-to-steel joint made by friction stir welding: A review. *Mater. Today Commun.* **2015**, *5*, 32–49. [CrossRef]

10. Mishra, R.S.; De, P.S.; Kumar, N. *Friction Stir Welding and Processing: Science and Engineering*; Springer International Publishing: Cham, Switzerland, 2014.
11. Kimapong, K.; Watanabe, T. Friction stir welding of aluminum alloy to steel. *Weld J.* **2004**, *83*, 277s–282s.
12. Elrefaey, A.; Gouda, M.; Takahashi, M.; Ikeuchi, K. Characterization of aluminum/steel lap joint by friction stir welding. *J. Mater. Eng. Perform.* **2005**, *14*, 10–17. [CrossRef]
13. Coelho, R.S.; Kostka, A.; Sheikhi, S.; Dos Santos, J.; Pyzalla, A.R. Microstructure and Mechanical Properties of an AA6181-T4 Aluminium Alloy to HC340LA High Strength Steel Friction Stir Overlap Weld. *Adv. Eng. Mater.* **2008**, *10*, 961–972. [CrossRef]
14. Wei, Y.; Li, J.; Xiong, J.; Zhang, F. Effect of tool pin insertion depth on friction stir lap welding of aluminum to stainless steel. *J. Mater. Eng. Perform.* **2013**, *22*, 3005–3013. [CrossRef]
15. Badarinarayan, H.; Shi, Y.; Li, X.; Okamoto, K. Effect of tool geometry on hook formation and static strength of friction stir spot welded aluminum 5754-O sheets. *Int. J. Mach. Tools Manuf.* **2009**, *49*, 814–823. [CrossRef]
16. Bozzi, S.; Helbert-Etter, A.L.; Baudin, T.; Criqui, B.; Kerbiguet, J.G. Intermetallic compounds in Al 6016/IF-steel friction stir spot welds. *Mater. Sci. Eng. A* **2010**, *527*, 4505–4509. [CrossRef]
17. Mendes, N.; Neto, P.; Loureiro, A.; Moreira, A.P. Machines and control systems for friction stir welding: A review. *Mater. Des.* **2016**, *90*, 256–265. [CrossRef]
18. Smith, C.B. Robotic friction stir welding using a standard industrial robot. In Proceedings of the Second Friction Stir Welding International Symposium, Gothenburg, Sweden, 27–29 June 2000.
19. Cook, G.E.; Crawford, R.; Clark, D.E.; Strauss, A.M. Robotic friction stir welding. *Ind. Robot Int. J.* **2004**, *31*, 55–63. [CrossRef]
20. Longhurst, W.R.; Strauss, A.M.; Cook, G.E.; Cox, C.D.; Hendricks, C.E.; Gibson, B.T.; Dawant, Y.S. Investigation of force-controlled friction stir welding for manufacturing and automation. *Proc. Inst. Mech. Eng. Pt. B J. Eng. Manuf.* **2009**, *224*, 937–949. [CrossRef]
21. Fehrenbacher, A.; Smith, C.B.; Duffie, N.A.; Ferrier, N.J.; Pfefferkorn, F.E.; Zinn, M.R. Combined Temperature and Force Control for Robotic Friction Stir Welding. *J. Manuf. Sci. Eng.* **2014**, *136*. [CrossRef]
22. Yoon, J.; Kim, C.; Rhee, S. Compensation of Vertical Position Error Using a Force–Deflection Model in Friction Stir Spot Welding. *Metals* **2018**, *8*, 1049. [CrossRef]
23. Movahedi, M.; Kokabi, A.H.; Seyed Reihani, S.M.; Cheng, W.J.; Wang, C.J. Effect of annealing treatment on joint strength of aluminum/steel friction stir lap weld. *Mater. Des.* **2013**, *44*, 487–492. [CrossRef]
24. Chen, Z.; Yazdanian, S. Friction Stir Lap Welding: Material flow, joint structure and strength. *J. Achiev. Mater. Manuf. Eng.* **2012**, *55*, 629–637.
25. Das, A.; Shome, M.; Das, C.R.; Goecke, S.-F.; De, A. Joining of galvannealed steel and aluminium alloy using controlled short circuiting gas metal arc welding process. *Sci. Technol. Weld. Join.* **2015**, *20*, 402–408. [CrossRef]

metals

MDPI

Article

Effect of Tool Rotational Speeds on the Microstructure and Mechanical Properties of a Dissimilar Friction-Stir-Welded CuCrZr/CuNiCrSi Butt Joint

Youqing Sun [1], Diqiu He [1,2], Fei Xue [2] and Ruilin Lai [1,*]

[1] State Key Laboratory of High Performance Complex Manufacturing, Central South University, Changsha 410083, China; sunyouqing@csu.edu.cn (Y.S.); 133701033@csu.edu.cn (D.H.)
[2] Light Alloy Research Institute, Central South University, Changsha 410083, China; syqcsu@csu.edu.cn
* Correspondence: hdqcsu@csu.edu.cn; Tel.: +86-0731-8887-6230

Received: 1 June 2018; Accepted: 4 July 2018; Published: 6 July 2018

Abstract: In this study, dissimilar CuNiCrSi and CuCrZr butt joints were friction stir welded at a constant welding speed of 150 mm/min, but at different rotational speeds of 800, 1100, 1400, 1700, and 2100 rpm. Sound joints were achieved at the rotational speeds of 1400 and 1700 rpm. It was found that the area of retreating material and grain size in the nugget zone increased with the increase of tool rotational speeds. The base metal on the CuNiCrSi side (CuNiCrSi-BM) contains a large density of Cr and δ-Ni_2Si precipitates, and a great deal of Cr precipitates can be observed in the base metal on the CuCrZr side (CuCrZr-BM). All these precipitates are completely dissolved into the matrix in both the nugget zone on the CuCrZr side (CuCrZr-NZ) and the nugget zone on the CuNiCrSi side (CuNiCrSi-NZ). The precipitation strengthening plays a dominant role in the base metals, but the grain boundary strengthening is more effective in improving the mechanical properties in the nugget zone. Both the hardness and tensile strength decrease sharply from the base metal to the nugget zone due to the dissolution of precipitates. Mechanical properties such as microhardness and tensile strength in the nugget zone decrease with the increase of rotational speeds because the grain size is larger at a higher rotational speed.

Keywords: dissimilar joints; friction-stir welding; the rotational speeds; microstructure; mechanical properties

1. Introduction

Friction-stir welding (FSW) is a solid-state joining process [1]. This technique is characterized by combining frictional heating and mechanical breakup arising from the rotating tool. Friction-stir welding was initially designed to weld aluminum alloys. However, with the improvement of this technique, FSW has been significantly expanded to the welding of high-melting-point materials, such as Cu and its alloys. The friction-stir-welded joints are controlled by several welding parameters. The tool rotational speed and the welding speed are two dominant welding parameters. Numerous studies have been carried out to study the influence of the tool rotational speed and the welding speed on the resulting properties of copper joints. Sun et al. [2] studied the microstructure and mechanical properties of FSW copper joints over a wide range of welding parameters, including rotational speed (from 750 to 1200 rpm) and welding speed (from 200 to 800 mm/min). They found that the grain size in the nugget zone (NZ) is much smaller than that of the base metal, and the highest tensile strength of joints can reach 380 MPa. Liu et al. [3] reported friction-stir-welded copper joints at different tool rotational speeds ranging from 300 to 1000 rpm. They found that the ultimate tensile strength (UTS) initially increased to the maximum (277 MPa) and then decreased as the rotational speeds increased from 300 to 1000 rpm. Azizi et al. [4] reported microstructure and mechanical properties of friction-stir-welded copper joints in plates with 10 mm thickness. They found that the ultimate tensile strength initially increases to a maximum (260 MPa) and then decreases when the welding speeds increase from 50 to 200 mm/min.

Dissimilar joints are currently of great interest in industrial applications due to their technical and economic benefits [5]. Copper and aluminum are the most common materials used in the dissimilar joints. Dissimilar joints of Al/Cu which combine copper's improved strength and electrical properties with aluminum's low weight are used widely in industrial fields. There are a number of studies which are related to dissimilar Al/Cu joints. Xue et al. [6] investigated the effects of the rotational speeds and tool offsets on dissimilar friction-stir-welded joints of 1060 aluminum and pure copper. They found that good tensile properties can be obtained at higher rotational speeds of 600–1000 rpm with a constant welding speed of 100 mm/min, as well as proper pin offsets of 2 and 2.5 mm to softer aluminium alloys. The maximum value of tensile strength in their study was 110 MPa. Tan et al. [7] discussed the microstructure and mechanical properties of dissimilar 5A02 aluminium and pure copper joints fabricated by FSW. They found that defect-free joints can be achieved under the condition of high rotational speeds (1100 rpm) and low welding speed (20 mm/min), with the tool offset by 0.2 mm relative to the weld centreline and Al sheet. They also found that sound joints can be obtained when the harder copper plate was fixed at the retreating side. Sahu et al. [8] systematically investigated the influence of welding parameters, including plate position, tool offsets, and tool rotational speeds, on the microstructure and mechanical properties of dissimilar 1051 aluminium and pure copper joints. They found that good mechanical properties can be obtained at a tool rotational speed of 1200 rpm, welding speed of 30 mm/min, 0.1 mm plunging depth, and 1.5 mm offset towards Al alloy. In this condition, the ultimate strength and the yield strength can respectively reach 126 MPa and 119.3 MPa.

Thus, the review of the existing literature exposes the substantial study on FSW of Cu and dissimilar Al/Cu joints. However, the conventional Cu joints and dissimilar Al/Cu joints possess limited capabilities to handle structural loads due to their relatively lower strength. It is also difficult for Al/Cu dissimilar joints to perform well in terms of electrical conductivity. The conventional Cu joints and dissimilar Al/Cu joints cannot fulfill the demands of critical functional and structural applications, which require both a high mechanical strength and a high electrical conductivity. CuCrZr and CuNiCrSi alloys, which are treated by solution and aging process, can possess a good combination of high strength and good electrical conductivities. The ultimate tensile strength and the electrical conductivity of the CuCrZr alloy can reach about 530 MPa and 80% international annealed copper standard (IACS) [9], respectively. In comparison with the CuCrZr alloy, the CuNiCrSi alloy possesses a higher ultimate tensile strength of 600–800 MPa, but a lower electrical conductivity of about 45% IACs, owing to the different additions of Ni and Si [10]. CuCrZr and CuNiCrSi alloys with both high strength and good electrical conductivity are in high demand in many industries, in such applications as large generator rotors and heat sink material for fusion reactor components [11,12]. Therefore, fabrication and processing technology of these alloys with both high strength and high conductivity are very important. Although substantial studies have been focused on the FSW of conventional copper alloys and dissimilar Al/Cu alloys, the reports concentrating on FSW of copper alloys with high strength and good electrical conductivity such as CuCrZr and CuNiCrSi alloys are limited. There are only a few studies on the FSW of CuCrZr. Sahlot et al. [13] discussed the wear of the tool used in the FSW of CuCrZr rather than characteristics of the joints. Jha et al. [14] studied the microstructure and mechanical properties of CuCrZr alloys welded by FSW. They found that the tensile strength of welded joints was lower than that of the base metal due to the dissolution of precipitates in the welded zone. Lai et al. [15] studied the microstructural properties of the CuCrZr joints welded by FSW. The thickness of CuCrZr plates in their study were 10 mm. They found that the grain size in the NZ was decreased gradually from the top to the bottom area of the welds due to the distinctive heat production and the heat dissipation on the welding joint, which cause the microhardness and tensile strength of the welds to be slightly increased from the top to the bottom area of the welds. To our knowledge, no studies have evaluated the effects of welding parameters on the properties of dissimilar CuCrZr and

CuNiCrSi joints fabricated by FSW. The present study systematically investigated the influence of tool rotational speeds on the microstructural evolution and mechanical properties of dissimilar CuCrZr and CuNiCrSi joints. The grain structure and precipitates of the investigated alloys after the FSW process were discussed in detail. The mechanical properties including microhardness and tensile strength were also studied in detail.

2. Materials and Methods

The base materials used in this study were rolled CuCrZr and CuNiCrSi plates with 3 mm thickness. All plates were cut into dimensions of 300 mm long and 100 mm wide before the welding process. The CuCrZr alloy was treated through solution (920 °C for 0.5 h) process and then aged at 440 °C for 2 h. The CuNiCrSi alloy was also subjected to solution (800 °C for 2 h) treatment followed by an aging (450 °C for 5 h) process. Table 1 shows the chemical compositions of these two alloys.

Table 1. The chemical compositions of the CuCrZr alloy and CuNiCrSi alloy.

Alloy (wt %)	Cu	Al	Mg	Cr	Ni	Zr	Fe	Si
CuCrZr	Bal.	0.25	0.1	0.8	-	0.3	0.09	0.04
CuNiCrSi	Bal.	-	-	0.5	2.0	-	≤0.15	0.5

2.1. Friction-Stir-Welding Process

Dissimilar CuCrZr and CuNiCrSi joints were friction-stir-welded by a tool at rotational speeds of 800, 1100, 1400, 1700, and 2000 rpm. The tool welding speed and the tool tilt angle were fixed at 150 mm/min and 2.5°, respectively. To investigate the effects of material positions on microstructure and mechanical properties, the CuCrZr and CuNiCrSi alloys were alternately placed on the advancing side (AS) and retreating side (RS), respectively. When CuNiCrSi was on the AS, then CuZrCr was on the retreating side (RS) of the welding tool pin. When CuZrCr was on the AS, then CuNiCrSi was on the RS. The detailed welding conditions are listed in Table 2. The friction-stir-welding process was performed on a specially constructed apparatus, which has been reported by our previous work [16]. The FSW tool was composed of a concave shoulder and a conical pin. The diameters of the shoulder and the length of the pin were 10 mm and 2.8 mm, respectively. The diameters of the pin were 3.5 mm at the root and 4.5 mm at the head. For the welding process, the tool rotated in the clockwise direction. The pin was slowly inserted into the workpieces with a constant plunging depth of 0.1 mm and plunging speed of 0.05 mm/s. The schematic presentation of the friction-stir-welding process and dimensions of the FSW tool are shown in Figure 1.

Table 2. The welding parameters used in the study.

Conditions	Material on the Advancing Side (AS)	Material on the Retreating Side (RS)	Rotational Speeds (rpm)	Travel Speed (mm/s)	Tilt Angle (°)
A1	CuNiCrSi	CuCrZr	800	150	2.5
A2	CuCrZr	CuNiCrSi	800	150	2.5
A3	CuNiCrSi	CuCrZr	1100	150	2.5
A4	CuCrZr	CuNiCrSi	1100	150	2.5
A5	CuNiCrSi	CuCrZr	1400	150	2.5
A6	CuCrZr	CuNiCrSi	1400	150	2.5
A7	CuNiCrSi	CuCrZr	1700	150	2.5
A8	CuCrZr	CuNiCrSi	1700	150	2.5
A9	CuNiCrSi	CuCrZr	2000	150	2.5
A10	CuCrZr	CuNiCrSi	2000	150	2.5

After the welding, a preliminary region of applicable rotational speeds was carefully chosen by eliminating joints which had groove-like defects and surface-galling defects on the surfaces of the joints. Then, X-ray radiography inspections were performed on an X-ray nondestructive testing system (XD7600NT, Dage, London, UK) to further reveal the weld defects in the inner zones of the welded joints. The welded joints were scanned along the weld line using a 100 KV X-ray source voltage.

Finally, joints were cut into strips perpendicularly to the welding line. Specimens for metallographic observation and mechanical testing were made from these strips.

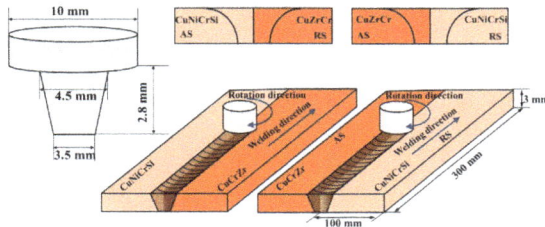

Figure 1. The schematic representation of the friction-stir-welding process and dimensions of the friction stir welding (FSW) tool.

2.2. Microstructural Characterization

The microstructures of samples under welding conditions were analyzed by optical microscope (OM), electron backscatter diffraction (EBSD), and transmission electron microscope (TEM). Specimens for OM analysis were polished according to a standard process and then etched with a 40 mL H_2O, 10 mL HCl, and 2 g Fe_3Cl solution. The transverse cross-section macrographs of the welded joints were observed by a 3D microscope (VHX 5000, KEYENCE, Osaka, Japan). For the EBSD analysis, the samples after mechanical polishing were further vibration-polished to remove stress. The grain structures of different zones including CuCrZr-BM, CuCrZr-NZ, CuNiCrSi-BM, and CuNiCrSi-NZ were analyzed using a FEI Quanta 650 FEG scanning electron microscopy (FEI Corporation, Hillsboro, OR, USA). For the TEM study, some thin foils of 0.5 mm thickness were cut perpendicular to the welding direction. Then, the foils were grinded into one of a thickness of 70 μm~80 μm and several (Φ3 mm) discs were punched out from different zones including the CuCrZr-BM, CuCrZr-NZ, CuNiCrSi-BM, and CuNiCrSi-NZ. A twin-jet electro-polisher was used to produce electron-transparent thin sections in these discs with a solution of 75% methanol and 25% nitric acid, using an electrolyte voltage of 10 V at −30 °C. TEM experiments were conducted on the Tecnai G2 F20 (FEI Corporation, Hillsboro, OR, USA) with an acceleration voltage of 120 keV.

2.3. Mechanical Testing

The tensile properties of the dissimilar joints were evaluated using a universal electronic tensile testing machine (MTS Systems Corporation, Eden Prairie, MN, USA). The tensile testing specimens with a gauge length of 150 mm and a width of 25 mm were machined perpendicularly to the welding direction using a wire electrical discharge machine (DK7720, Terui, Taizhou, China). The tensile tests were performed three times for each welding condition with a testing speed of 2 mm/min at room temperature. The measurement of Vickers hardness was conducted along the centerline, using a Vickers hardness machine (Huayin Testing Instrument Co., Ltd., Yantai, China) with a load of 100 g and a dwell time of 10 s. The distance between each neighboring measured points was 0.5 mm. The Vickers hardness tests were repeated three times under each welding condition to obtain the average microhardness of welds.

3. Results

3.1. Surface Morphologies and X-ray Radiographs of the Joints

Figure 2 shows the surface morphologies of dissimilar CuCrZr/CuNiCrSi butt joints under conditions A1, A2, A9, and A10. It is seen that the groove-like defects are formed on the AS side at the lower rotational speeds of 800 rpm. However, surface galling defects are seen to occur at the higher rotational speed of 2000 rpm.

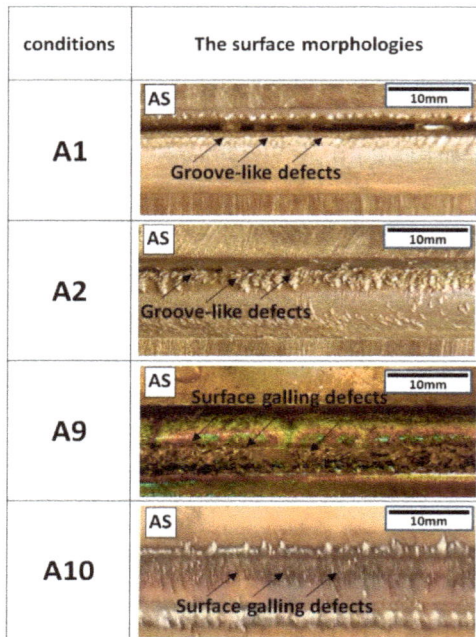

Figure 2. The surface morphologies of dissimilar CuCrZr/CuNiCrSi butt joints under conditions A1, A2, A9, and A10.

Figure 3 shows the surface morphologies and the relevant X-ray radiographs of welded joints under conditions A3–A8. When the rotational speed is 1100 rpm, tunnelling defects are found to be formed inside the joints by X-ray radiographs, but these defects cannot be seen on the surface morphologies. The defect-free joints are formed at the rotational speeds of 1400 rpm and 1700 rpm. Based on the surface morphologies and X-ray radiographs of the stir zone, it can be concluded that the rotational speeds of 1400 rpm and 1700 rpm are adequate rotational speeds for dissimilar friction-stir-welded CuCrZr/CuNiCrSi butt joints at the constant travel speed of 150 mm/min.

3.2. Microstructure of Dissimilar CuCrZr/CuNiCrSi Butt Joints

Figure 4 shows the transverse cross-section macrographs of the welded joints obtained under conditions A3–A8. Due to the difference in etching response, the CuNiCrSi alloy appears as a light color, whereas the CuCrZr alloy appears as the dark colour regions. When the rotational speed is 1100 rpm, a tunnelling defect is formed at the bottom of the AS side with the CuNiCrSi alloy located on the AS, while two smaller defects are observed in the NZ near the AS with the inverse material positions. Sound joints are produced at the rotational speeds of 1400 and 1700 rpm. The cross-section micrographs are in complete conformity with the X-ray radiographs seen in Figure 3. Some discernable differences can be found between these cross-section macrographs. Two stir patterns can be identified from these differences: one is that the area of retreating materials in the NZ increases with the increase of rotational speeds; the other is that the area of retreating materials in the NZ is seen to be a little bigger when the CuNiCrSi alloy is placed on the RS.

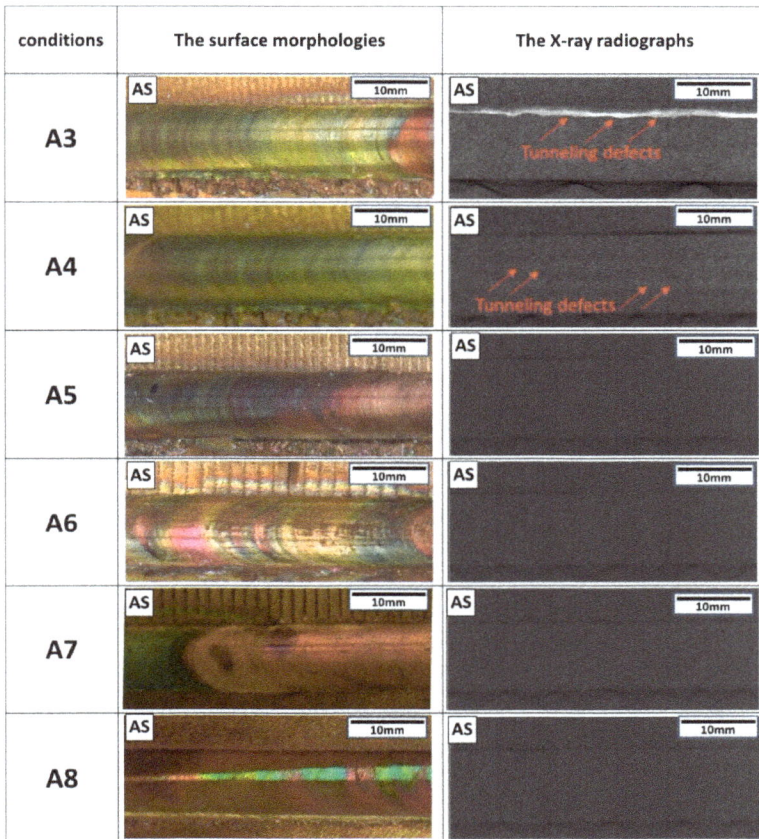

Figure 3. The surface morphologies and the relevant X-ray radiographs of welded joints under conditions A3–A8.

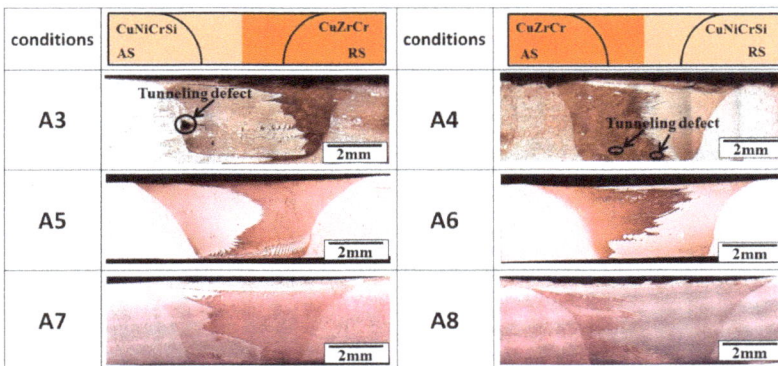

Figure 4. The transverse cross-section macrographs of the welded joints obtained under conditions A3–A8.

Figure 5 shows inverse pole figure maps in the different regions, including the CuCrZr-BM, CuNiCrSi-BM, CuCrZr-NZ, and CuNiCrSi-NZ. The grain size frequencies of different zones under different rotational speeds of 1400 and 1700 rpm are also shown in Figure 6. Figure 5a,b shows that both the CuNiCrSi-BM and CuCrZr-BM exhibit a rolled structure with coarse grains of different size. It can be found from Figure 6a,b that the average grain sizes of the CuNiCrSi-BM and CuCrZr-BM are 33.30 µm and 40.53 µm, respectively. In comparison, the CuNiCrSi-NZ (Figure 5c,e) and the CuCrZr-NZ (Figure 5d,f) are composed of equiaxed grains with even distribution. Moreover, the grain size in the NZ is observed to increase with the increase of rotational speeds. The average grain sizes of the CuNiCrSi-NZ and CuCrZr-NZ are 0.95 µm (Figure 6c) and 1.42 µm (Figure 6d) at the rotational speed of 1400 rpm, respectively. However, the average grain sizes of the CuNiCrSi-NZ and CuCrZr-NZ increase to 1.48 µm (Figure 6e) and 2.32 µm (Figure 6f) with the rotational speed increased to 1700 rpm.

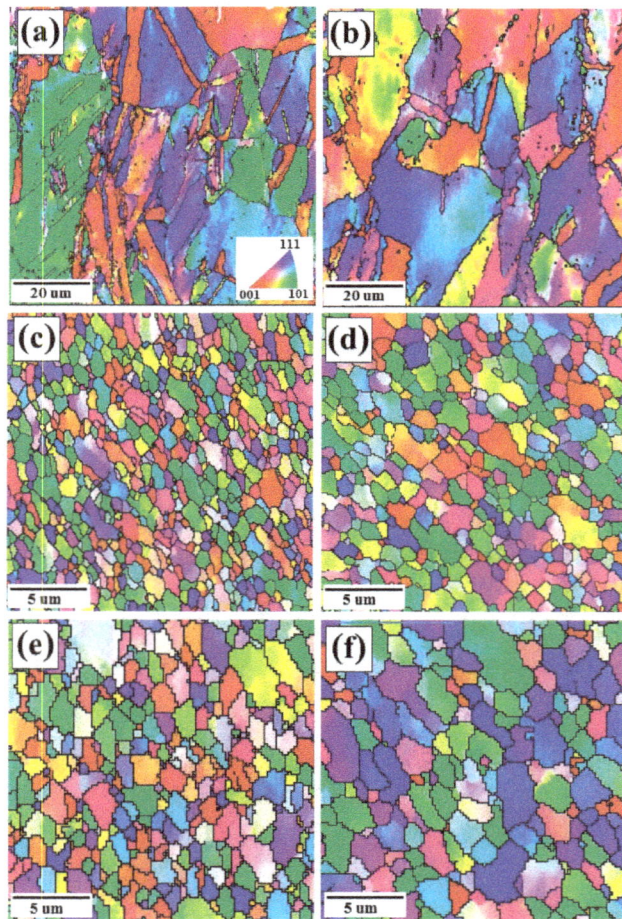

Figure 5. Inverse pole figure maps in the different regions. (**a**) CuNiCrSi-BM; (**b**) CuCrZr-BM; (**c**) CuNiCrSi-NZ under condition A5; (**d**) CuCrZr-NZ under condition A5; (**e**) CuNiCrSi-NZ under condition A7; (**f**) CuCrZr-NZ under condition A7.

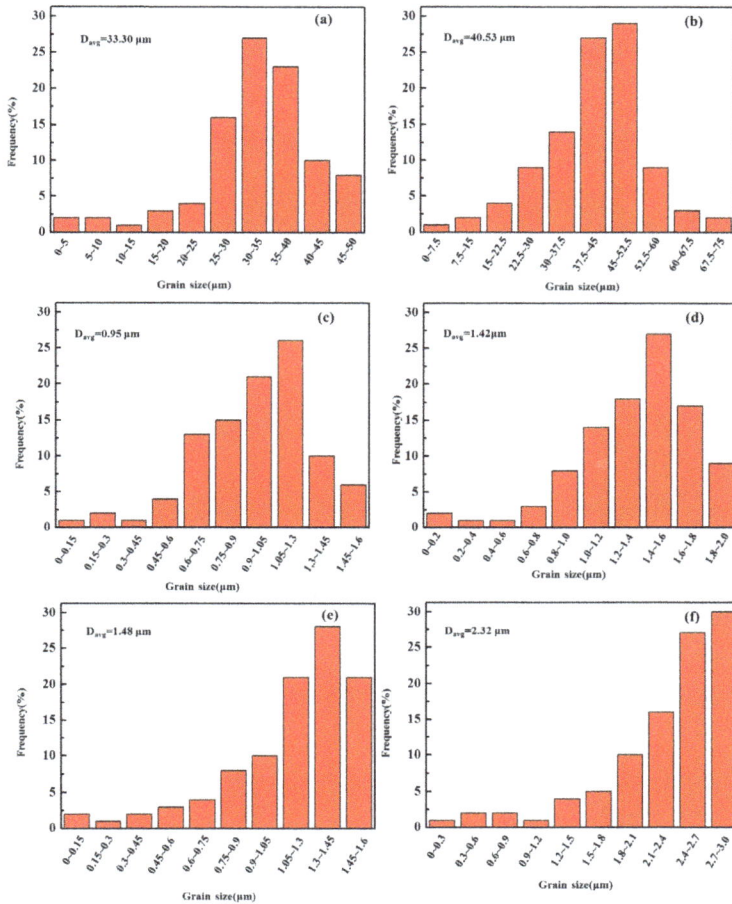

Figure 6. Grain size frequencies of different regions. (**a**) CuNiCrSi-BM; (**b**) CuCrZr-BM; (**c**) CuNiCrSi-NZ under condition A5; (**d**) CuCrZr-NZ under condition A5; (**e**) CuNiCrSi-NZ under condition A7; (**f**) CuCrZr-NZ under condition A7. (D_{avg} means the average diameter of grains).

Figure 7 shows the bright-field TEM micrographs of samples from different zones at the rotational speed of 1400 rpm (condition A5). The relevant selected-area electron diffraction (SAD) patterns are shown in the top left-hand corner of each TEM micrograph as well. Figure 7a shows that the lobe-lobe contrast precipitates are scattered randomly in the CuCrZr-BM. The average length of these precipitates is 5–8 nm. In addition, the relevant SAD pattern parallel to the $[011]_{Cu}$ direction reveals the reflections spots from the Cu matrix and precipitates. Moreover, the face center cubic (FCC) Cr precipitates can be identified as the contributors to the reflections spots of the precipitates. In fact, these Cr precipitates are found to be generated in the aging process of the CuCrZr alloy due to the solubility limit of the Cr element in copper [17], and these Cr precipitates can mitigate the movement of dislocations so as to improve the strengths of the CuCrZr alloy. Figure 7c shows that the lobe-lobe contrast Cr precipitates can also be detected in the CuNiCrSi-BM. However, these Cr precipitates cannot be identified by the reflection spots from the relevant SAD pattern parallel to $[011]_{Cu}$. Apart from Cr precipitates, another type of precipitates which are rod-shaped in the $[011]_{Cu}$ direction (Figure 7c) and disc-shaped in the

[111]$_{Cu}$ direction (Figure 7d) can be found in the CuNiCrSi-BM. Moreover, the corresponding SAD patterns parallel to the [011]$_{Cu}$ direction and the [111]$_{Cu}$ direction supported the assumption that the δ-Ni$_2$Si precipitates existed in the CuNiCrSi alloy. Similar results can also be found in previous works about the CuNiCrSi alloy [18,19]. After the welding, Figure 7b,e shows that all precipitates are dissolved into the matrix in both the CuCrZr-NZ and CuNiCrSi-NZ.

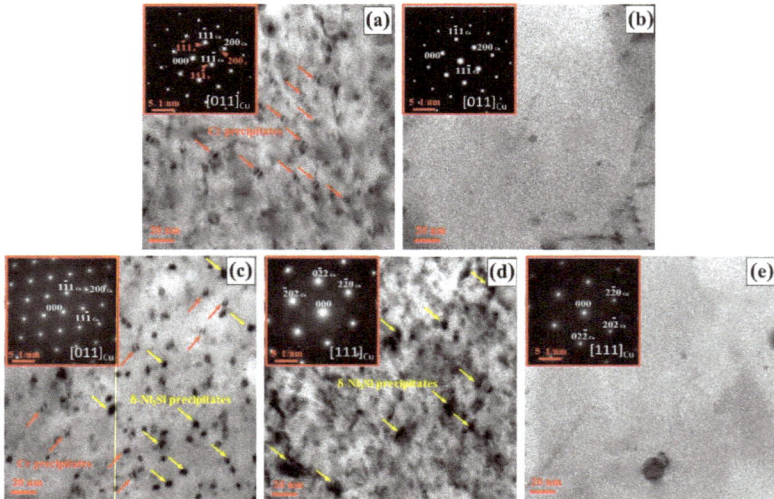

Figure 7. The bright-field TEM micrographs of samples from different zones at the rotational speed of 1400 rpm (condition A5). (**a**) CuCrZr-BM in the [011]$_{Cu}$ direction; (**b**) CuCrZr-NZ in the [011]$_{Cu}$ direction; (**b**) CuNiCrSi-BM in the [011]$_{Cu}$ direction; (**d**) CuNiCrSi-BM in the [111]$_{Cu}$ direction; (**e**) CuNiCrSi-NZ in the [111]$_{Cu}$ direction.

Figure 8 shows the TEM micrographs of samples from the CuCrZr-NZ and CuNiCrSi-NZ when the rotational speed increased to 1700 rpm (condition A7). As shown in Figure 7a,b, all precipitates are dissolved into the Cu matrix in the NZ, which is similar to the results under the rotational speed of 1400 rpm.

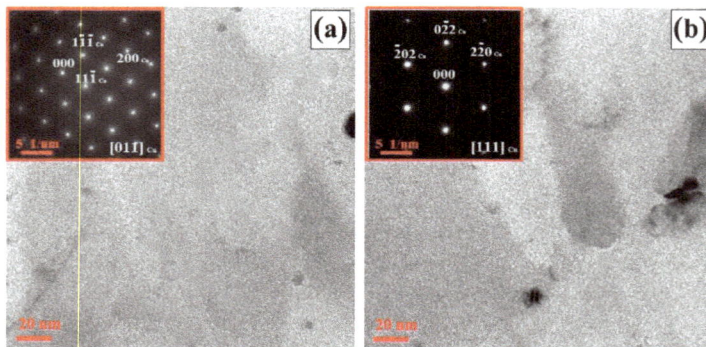

Figure 8. The bright-field TEM micrographs of samples from the CuCrZr-NZ and CuNiCrSi-NZ when the rotational speed increased to 1700 rpm (condition A7). (**a**) CuCrZr-NZ in the [011]$_{Cu}$ direction; (**b**) CuNiCrSi-NZ in the [111]$_{Cu}$ direction.

3.3. Mechanical Properties of CuCrZr/CuNiCrSi Butt Joints

Figure 9 illustrates the microhardness along the centreline of dissimilar CuCrZr/CuNiCrSi joints produced under different rotational speeds (conditions A5–A8). The distribution of the microhardness profile is asymmetrical along the measuring line, which is caused by different mechanical properties of the CuCrZr and CuNiCrSi alloys. The CuNiCrSi-BM and CuCrZr-BM demonstrate average microhardness values of about 225 HV and 155 HV, respectively. However, both the CuNiCrSi-NZ and CuCrZr-NZ show lower hardness relative to the base metal. The average microhardness in the CuNiCrSi-NZ and CuCrZr-NZ is 150 HV and 125 HV, respectively. The CuCrZr-NZ is the softest region of the whole dissimilar joint. In addition, it can be seen from Figure 9 that the microhardness of the NZ decreases with the increase of rotational speeds.

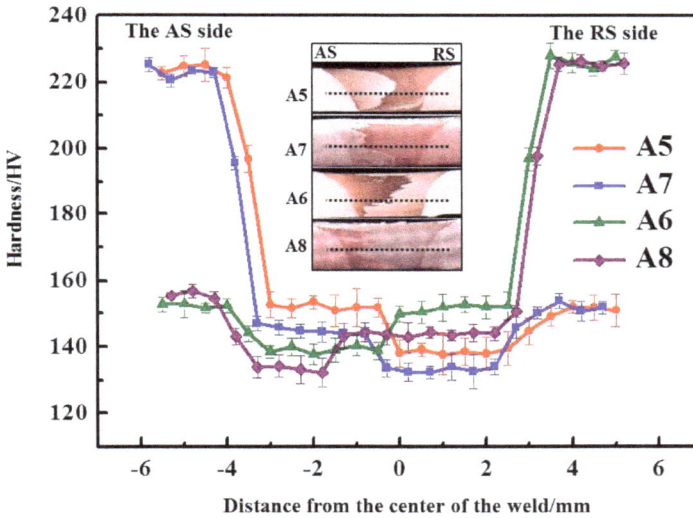

Figure 9. The microhardness along the centreline of dissimilar CuCrZr/CuNiCrSi joints.

Figure 10 demonstrates the mechanical properties of dissimilar joints. Table 3 describes the ultimate tensile strength, yield strength, and elongation of different joints in detail. Figure 10a shows the variation tendency of mechanical properties when the CuCrZr alloy is located on the AS. It is seen that both the tensile strength and the yield strength tend to increase first and then decrease slightly as the rotational speeds increase from 1100 rpm to 1700 rpm. Figure 10b shows that the same variation tendency is achieved in dissimilar joints when the CuNiCrSi alloy is located on the AS. Figure 10c shows that the strengths of welds are affected by the material position at the same rotational speed. The tensile strength of joints with CuCrZr located on the AS is slightly greater than when CuCrZr is on the RS. The maximum value of tensile strength is about 445 MPa in all welds, which is found to be only 80% and 60% of the tensile strengths of the CuCrZr and CuNiCrSi alloys, respectively. However, the strength of the joints obtained in the present study is still higher than that of most friction-stir-welded copper joints mentioned in previous studies [4,15].

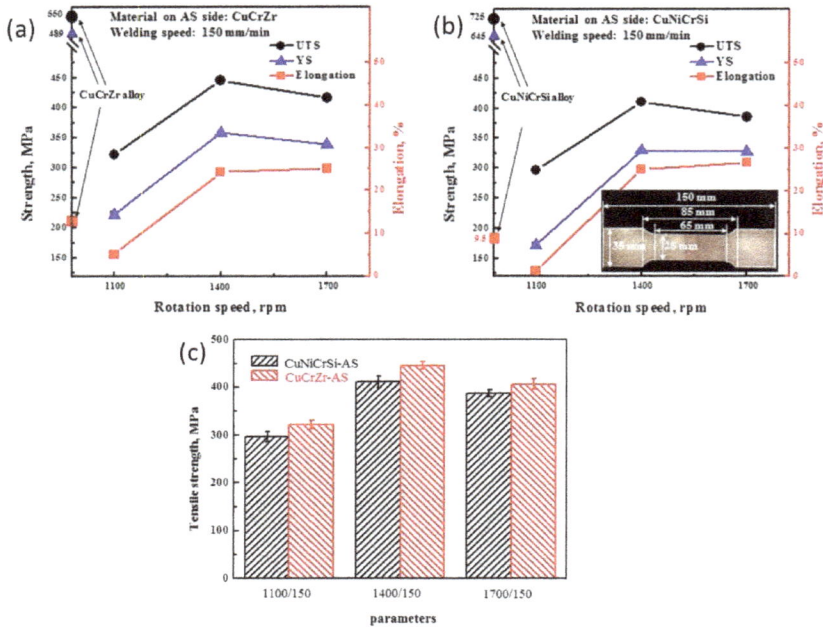

Figure 10. The mechanical properties of dissimilar joints. (**a**) The variation tendency of mechanical properties when the CuCrZr alloy is located on the AS; (**b**) the variation tendency of mechanical properties when the CuNiCrSi alloy is located on the AS; (**c**) tensile strength of joints. (UTS means the ultimate tensile strength; YS means the yield strength).

Table 3. The ultimate tensile strength, yield strength, and elongation of different joints.

Conditions	Material on AS Side	UTS (MPa)	YS (MPa)	Elongation (%)
A3	CuNiCrSi	296.82 ± 10	171.73 ± 5	1.32 ± 0.5
A4	CuCrZr	321.77 ± 8	221.59 ± 5	5.1 ± 0.9
A5	CuNiCrSi	410.76 ± 12	329.82 ± 3	25.02 ± 1.2
A6	CuCrZr	445.56 ± 9	358.38 ± 7	24.25 ± 0.8
A7	CuNiCrSi	386.09 ± 7	318.08 ± 6	26.62 ± 1.5
A8	CuCrZr	405.78 ± 11	338.51 ± 10	25.13 ± 1.3
CuNiCrSi	-	725 ± 13	646 ± 9	9.5 ± 1.5
CuCrZr	-	550 ± 15	489 ± 10	11.5 ± 1.0

Figure 11 shows the failure locations of tensile testing specimens under the different conditions A3–A8. Obviously, all the welds failed at the stir zone in spite of different rotational speeds. The tensile specimens failed directly at the position of tunnelling defects with no necking under conditions A3 and A4. In comparison, other welds without any defects exhibit some apparent necking in the process of tensile testing. It is worth noting that the relative material position exerts an influence on the failure locations of the tensile specimens, although the previous studies of dissimilar joints always ignored this issue. Specifically, it is seen that the fracture location of the tensile specimens is the CuCrZr-NZ with the CuNiCrSi alloy on the AS (conditions A5 and A7), but the tensile specimens failed at the mixed zone containing the CuNiCrSi and CuCrZr alloys when the CuNiCrSi-NZ and CuCrZr-NZ are in the inverse material position (conditions A6 and A8).

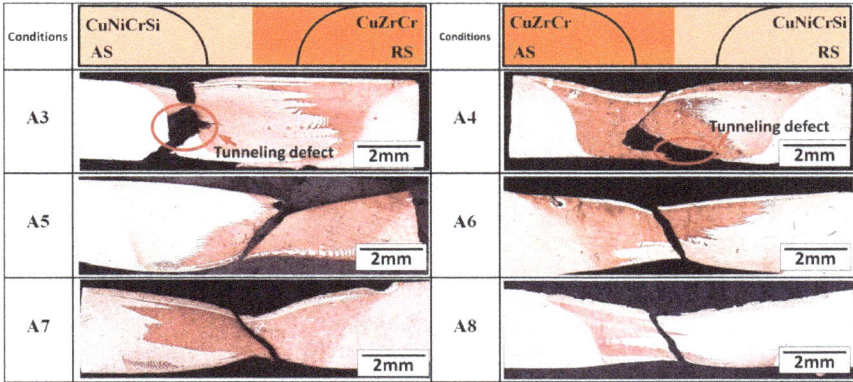

Figure 11. The failure locations of tensile testing specimens under the different conditions A3–A8.

Figure 12 shows the SEM micrographs of the fracture surface of the dissimilar joints under conditions A5 and A7, respectively. Fine populated dimples were observed on fracture surfaces of joints under conditions A5 and A7, which indicates that the failure mode is ductile fracture in these two conditions. There is seen to be no significant difference in the fracture mechanism under conditions A5 and A7. Both selected samples of tensile testing specimens experience extensive plastic deformation during the process of failure in these two conditions.

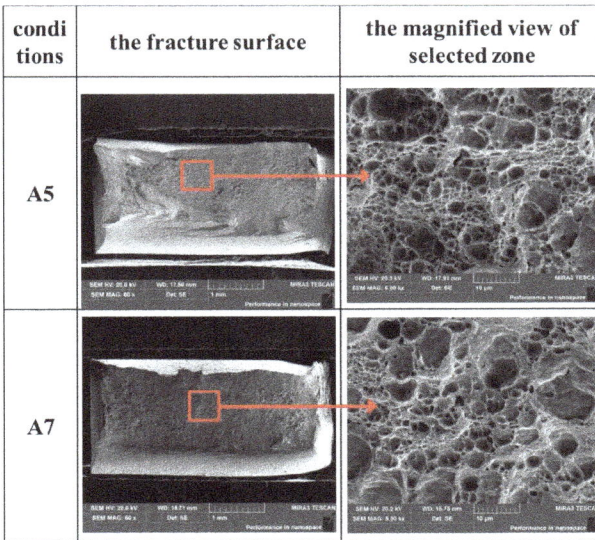

Figure 12. The SEM micrographs of the fracture surface of the dissimilar joints under conditions A5 and A7.

4. Discussion

4.1. Effect of Rotational Speeds on the Formation of Welds

In Figures 2 and 3, it can be seen that tunneling defects and groove-like defects occur at lower rotational speeds, while higher rotational speed produces galling defects on the surface of joints. The difference between the tunnelling defects and groove-like defects is that tunnelling defects cannot be found on the surface of joints. It is well documented that the heat input resulting from the rotational tool is the key factor in the formation of defects during FSW. The rotational speed has a direct relationship with heat input. According to Equation (1) [20], the maximum temperature during FSW increases with the increase of tool rotational speeds when at a constant tool welding speed. When at the lower rotational speeds of 800 rpm and 1100 rpm, the induced heat is insufficient, which results in less material softening and low plastic flow [21]. Therefore, the groove-like defects and tunnelling defects take place in these conditions. Ajri and Shin [22] offered a numerical model whereby to predict the formation of defects during FSW. The tunnelling defects are formed under the condition that the pin-induced flow does not drive the material on the AS to the RS, while the shoulder-induced flow does. In contrast, groove-like defects occurred when both the shoulder- and pin-induced flow fail to move the material from the advancing side to retreating side. Therefore, tunnelling defects are not visible on the top surface, but are present at the bottom of the joints. The results of the present study are highly consistent with this numerical modelling. However, when the rotational speed increases to 2100 rpm, the surface of the joint appears to contain surface galling as a result of too much heat input during the FSW [23].

$$\frac{T}{T_{\mathrm{m}}} = K \left(\frac{w^2}{v \times 10^4} \right)^{\alpha},$$ (1)

where T is the welding temperature (°C), T_{m} is the melting temperature of the plates (°C), $0.04 < \alpha < 0.06$ and $0.65 < K < 0.75$ are two defined constants, and w and v are the rotational and welding speeds, respectively.

4.2. Effects of Rotational Speeds on the Microstructure of Joints

Two stir patterns can be identified from the cross-section microstructures when the rotational speeds range from 1100 rpm to 1700 rpm. The key point of the difference between these two stir patterns is the area of retreating materials in the NZ. It is shown in Figure 4 that the area of the retreating section of material in the stir zone increases with the increasing of rotational speeds, which is mainly influenced by the material flow at different rotational speeds. Zhu et al. [24] indicated that the material flow velocity is reduced due to the insufficient driving force when the material on the AS was moved to the RS along the circular path. Moreover, the flow velocity and friction force of material on the rear AS reached minimum values. All these resulted in difficulties for the material to move from the RS to the AS. At higher rotational speeds, the material in the stir zone is softened enough that the material flow and the friction force are powerful enough to move it. Thus, a bigger part of the retreating section of material in the stir zone is pushed to the AS. However, this effect seems to be weakened when the CuNiCrSi plate is located on the AS, because the area of the retreating section of materials seems a little smaller than that when CuNiCrSi alloy is located on the AS. This is mainly because of the higher flow stress of the harder CuNiCrSi alloy. In our previous work [16], we found that the softer CuCrZr alloy is more difficult to push to the AS by the rotational tool due to the bigger resistance caused by the CuNiCrSi alloy when the CuNiCrSi plate is placed on the AS. In addition to that, both the velocity and friction force of the CuCrZr alloy were reduced as it is moved to the AS. In comparison, since it is easier for CuNiCrSi to move to the AS, the area of retreating materials seems to be larger in the NZ when the CuNiCrSi plate is located on the RS. Similar results can be found in other dissimilar friction-stir-welded joints, such as dissimilar joints of AA6061 and AA7075 [25]. Investigation found that the material flow is more difficult when the harder AA7075 was located on the AS.

Figures 5 and 6 show that the grain size increases with the increase of rotational speeds. This phenomenon can be explained by grain growth during the dynamic recrystallization. As is well-known, the rotational tool used in the FSW provides the material in the NZ with the frictional heating and the plastic flow. Then, the softened material is forced to rotate along the circular path by the rotational tool. In this process, dynamic recrystallization occurs due to the deformation at a high temperature. Consequently, the equiaxed grains in the NZ experience nucleation and then grain growth. So, it can be confirmed that the deformation and temperature are the key factors to the generation of the dynamic recrystallization and recrystallized grains; that is to say, recrystallized grain size in the NZ is mainly dependent on two factors: the peak temperature and the degree of deformation [3]. When the rotational speed increases, both the degree of deformation and the peak temperature increase. Increasing peak temperature causes larger grain size, but the increase of the degree of deformation has the opposite effect. In the present study, the grain size increases as the rotational speeds increase. Thus, it is seen that the dominant factor which influences the recrystallization phenomena in this study is the peak temperature. In fact, the higher peak temperature, which means a larger heat input, can provide more energy for grain growth.

In Figures 7 and 8, precipitates distributing in the BM are dissolved into the matrix in the NZ when the rotational speeds are 1400 rpm and 1700 rpm. The high welding heat produced by the frictional work and material deformation during FSW is responsible for the dissolution of nano-level strengthening precipitates in the NZ. In general, the heat input during the friction stir welding of Cu joints is usually high as a result of the high thermal conductivity of copper alloy. Jha et al. [14] found that the peak temperature can reach over 800 °C in the friction stir welding of aged CuCrZr plates. A peak temperature which is higher than the solvus temperature of precipitates can produce a supersaturated solution condition, which results in the dissolution of strengthening precipitates into the Cu matrix in the NZ. Similar results can also be found in the friction-stir-welding process of other precipitate-hardening alloys, such as 6063 aluminium [26], 7075 aluminium [27], and thick CuCrZr plates [15].

4.3. Effects of Rotational Speeds on the Mechanical Properties of Joints

In Figures 9 and 10, both the hardness and tensile strength decreases sharply from the BM to the NZ. In addition, tensile strength and the hardness of the NZ decrease slightly with the rotational speed increasing from 1400 rpm to 1700 rpm. Precipitation strengthening and grain boundary strengthening are important strengthening mechanisms of the studied alloy. The effects of precipitation strengthening are associated with the size and the density of precipitates. Small size and large density of precipitates can more effectively impede the dislocation movement and then strengthen the alloys [28]. The grain boundary strengthening is related to the size of grains. Small grain size can create a high density of grain boundaries that hinders the movement of dislocation, thereby improving the mechanical properties of alloys [28].

(1) Regarding the strengthening mechanisms of the BM, the grain boundary strengthening is limited due to the large grain size in the BM, which can reach up to 30–50 μm. However, the CuCrZr-BM contains a large density of Cr precipitates, while a great deal of Cr and δ-Ni_2Si precipitates can be detected in the CuNiCrSi-BM, which can contribute a strong precipitation strengthening effect by hindering the movements of dislocations.

(2) For the strengthening mechanisms of the NZ, on the one hand, all precipitates distributed in the BM are dissolved into the matrix in the NZ. The precipitation strengthening cannot work in the NZ. On the other hand, the grain size in the NZ is small when compared with that in the BM. In this case, the grain boundary strengthening plays a dominant impact on the mechanical properties in the NZ. Because the grain size in the NZ increases with the increasing of the rotational speeds, mechanical properties such as microhardness and tensile strength in the NZ decrease when the rotational speed increases from 1400 rpm to 1700 rpm.

In Figure 11, the tensile specimens all failed at the CuCrZr-NZ with the CuCrZr alloy located on the RS side. This is because the microhardness is lowest in the CuCrZr-NZ, which means the CuCrZr-NZ is the softest zone in the whole joint. However, when the CuCrZr alloy is located at the AS, the CuNiCrSi alloy at the RS is taken to the AS easily, so the specimens failed at the mixed zone of the CuCrZr-NZ and CuNiCrSi-NZ in the NZ. In this case, the tensile strength is a little higher than that when the CuNiCrSi alloy is located at the RS.

5. Conclusions

In the present study, the dissimilar CuNiCrSi and CuCrZr joints were friction-stir-welded at a constant welding speed of 150 mm/min and various rotational speeds of 800, 1100, 1400, 1700, and 2100 rpm, and the effects of rotational speeds on the microstructure and mechanical properties of the dissimilar CuNiCrSi and CuCrZr joints were investigated. The following conclusions can be drawn:

(I)   Dissimilar joints without any defects are obtained at rotational speeds of 1400 and 1700 rpm. Groove-like defects and tunneling defects are formed along the weld line at the lower rotational speeds of 800 and 1100 rpm. However, surface-galling defects are seen to occur at the higher rotational speed of 2100 rpm.

(II) The area of retreating materials and the grain size in the NZ increases with the increasing of rotational speeds. The CuNiCrSi-BM contains a large density of Cr and δ-Ni$_2$Si precipitates, while a great deal of Cr precipitates is detected in the CuCrZr-BM. All these precipitates are completely dissolved into the NZ as a consequence of high welding speed.

(III) Precipitation strengthening plays a dominant role in the BM. Both hardness and tensile strength decrease sharply from the BM to the NZ due to the dissolution of precipitates. Grain boundary strengthening plays a dominant impact on the mechanical properties in the NZ. Mechanical properties such as microhardness and tensile strength in the NZ decrease with the rotational speed increasing.

(IV) The CuCrZr-NZ is the softest zone in the whole joint. The fracture location of the tensile specimens is the CuCrZr-NZ with the CuNiCrSi alloy fixed on the AS, but the tensile specimens failed at the mixed zone of the two alloys when CuNiCrSi was on the RS.

Author Contributions: D.H. was the principle investigator of the research. Y.S., F.X., and R.L. carried out the welding tests and characterized the microstructure of the welded samples. Y.S. performed mechanical tests of hardness, fractography, and wrote the paper.

Funding: This research was funded by the National Basic Research Program of China ("973 Program", 2014CB046605).

Acknowledgments: This work was supported by Science and Technology Innovation Projects of graduate students of central south university (2017zzts653) and the National Basic Research Program of China ("973 Program", 2014CB046605). Youqing Sun especially wishes to thank Shu Li, for offering encouragement and help during the process of research.

Conflicts of Interest: The authors declare no conflict of interest.

References

1.   Ahmed, M.M.Z.; Ataya, S.; El-Sayed Seleman, M.M.; Ammar, H.R.; Ahmed, E. Friction stir welding of similar and dissimilar AA7075 and AA5083. *J. Mater. Process. Technol.* **2017**, *242*, 77–91. [CrossRef]
2.   Sun, Y.F.; Fujii, H. Investigation of the welding parameter dependent microstructure and mechanical properties of friction stir welded pure copper. *Mater. Sci. Eng. A* **2010**, *527*, 6879–6886. [CrossRef]
3.   Liu, H.J.; Shen, J.J.; Huang, Y.X.; Kuang, L.Y.; Liu, C.; Li, C. Effect of tool rotation rate on microstructure and mechanical properties of friction stir welded copper. *Sci. Technol. Weld. Join.* **2009**, *14*, 577–583. [CrossRef]
4.   Azizi, A.; Barenji, R.V.; Barenji, A.V.; Hashemipour, M. Microstructure and mechanical properties of friction stir welded thick pure copper plates. *Int. J. Adv. Manuf. Technol.* **2016**, *86*, 1985–1995. [CrossRef]
5.   Zhang, Q.Z.; Gong, W.B.; Liu, W. Microstructure and mechanical properties of dissimilar Al-Cu joints by friction stir welding. *Trans. Nonferr. Met. Soc. China* **2015**, *25*, 1779–1786. [CrossRef]

6. Xue, P.; Ni, D.R.; Wang, D.; Xiao, B.L.; Ma, Z.Y. Effect of friction stir welding parameters on the microstructure and mechanical properties of the dissimilar Al–Cu joints. *Mater. Sci. Eng. A* **2011**, *528*, 4683–4689. [CrossRef]

7. Tan, C.W.; Jiang, Z.G.; Li, L.Q.; Chen, Y.B.; Chen, X.Y. Microstructural evolution and mechanical properties of dissimilar Al–Cu joints produced by friction stir welding. *Mater. Des.* **2013**, *51*, 466–473. [CrossRef]

8. Sahu, P.K.; Pal, S.; Pal, S.K.; Jain, R. Influence of plate position, tool offset and tool rotational speed on mechanical properties and microstructures of dissimilar Al/Cu friction stir welding joints. *J. Mater. Process. Technol.* **2016**, *235*, 55–67. [CrossRef]

9. Mishnev, R.; Shakhova, I.; Belyakov, A.; Kaibyshev, R. Deformation microstructures, strengthening mechanisms, and electrical conductivity in a Cu–Cr–Zr alloy. *Mater. Sci. Eng. A* **2015**, *629*, 29–40. [CrossRef]

10. Gholami, M.; Vesely, J.; Altenberger, I.; Kuhn, H.A.; Janecek, M.; Wollmann, M.; Wagner, L. Effects of microstructure on mechanical properties of CuNiSi alloys. *J. Alloys Compd.* **2017**, *696*, 201–212. [CrossRef]

11. Shueh, C.; Chan, C.K.; Chang, C.C.; Sheng, I.C. Investigation of vacuum properties of CuCrZr alloy for high-heat-load absorber. *Nucl. Instrum. Methods Phys. Res.* **2017**, *841*, 1–4. [CrossRef]

12. Lipa, M.; Durocher, A.; Tivey, R.; Huber, T.; Schedler, B.; Weigert, J. The use of copper alloy CuCrZr as a structural material for actively cooled plasma facing and in vessel components. *Fusion Eng. Des.* **2005**, *75*, 469–473. [CrossRef]

13. Sahlot, P.; Jha, K.; Dey, G.K.; Arora, A. Quantitative wear analysis of H13 steel tool during friction stir welding of Cu-0.8%Cr-0.1%Zr alloy. *Wear* **2017**, *378–379*, 82–89. [CrossRef]

14. Jha, K.; Kumar, S.; Nachiket, K.; Bhanumurthy, K.; Dey, G.K. Friction stir welding (FSW) of aged CuCrZr alloy plates. *Metall. Mater. Trans. A* **2018**, *49*, 223–234. [CrossRef]

15. Lai, R.; He, D.; He, G.; Lin, J.; Sun, Y. Study of the microstructure evolution and properties response of a friction-stir-welded copper-chromium-zirconium alloy. *Metals* **2017**, *7*, 381. [CrossRef]

16. Sun, Y.; He, D.; Xue, F.; Lai, R.; He, G. Microstructure and mechanical characterization of a dissimilar friction-stir-welded CuCrZr/CuNiCrSi butt joint. *Metals* **2018**, *8*, 325. [CrossRef]

17. Holzwarth, U.; Stamm, H. The precipitation behaviour of ITER-grade Cu–Cr–Zr alloy after simulating the thermal cycle of hot isostatic pressing. *J. Nucl. Mater.* **2000**, *279*, 31–45. [CrossRef]

18. Lei, Q.; Xiao, Z.; Hu, W.; Derby, B.; Li, Z. Phase transformation behaviors and properties of a high strength Cu-Ni-Si alloy. *Mater. Sci. Eng. A* **2017**, *697*, 37–47. [CrossRef]

19. Lockyer, S.A.; Noble, F.W. Precipitate structure in a Cu-Ni-Si alloy. *J. Mater. Sci.* **1994**, *29*, 218–226. [CrossRef]

20. Mishra, R.S.; Ma, Z.Y. Friction stir welding and processing. *Mater. Sci. Eng. R* **2005**, *50*, 1–78. [CrossRef]

21. Zoeram, A.S.; Anijdan, S.H.M.; Jafarian, H.R.; Bhattacharjee, T. Welding parameters analysis and microstructural evolution of dissimilar joints in Al/Bronze processed by friction stir welding and their effect on engineering tensile behavior. *Mater. Sci. Eng. A* **2017**, *687*, 288–297. [CrossRef]

22. Ajri, A.; Shin, Y.C. Investigation on the effects of process parameters on defect formation in friction stir welded samples via predictive numerical modeling and experiments. *J. Manuf. Sci. Eng.* **2017**, *139*, 111009. [CrossRef]

23. Zettler, R.; Vugrin, T.; Schmücker, M. Effects and defects of friction stir welds. *Frict. Stir Weld.* **2010**, *23*, 245–276.

24. Zhu, Y.; Chen, G.; Chen, Q.; Zhang, G.; Shi, Q. Simulation of material plastic flow driven by non-uniform friction force during friction stir welding and related defect prediction. *Mater. Des.* **2016**, *108*, 400–410. [CrossRef]

25. Guo, J.F.; Chen, H.C.; Sun, C.N.; Bi, G.; Sun, Z.; Wei, J. Friction stir welding of dissimilar materials between AA6061 and AA7075 Al alloys effects of process parameters. *Mater. Des.* **2014**, *56*, 185–192. [CrossRef]

26. Sato, Y.S.; Kokawa, H.; Enomoto, M.; Jogan, S. Microstructural evolution of 6063 aluminum during friction-stir welding. *Metall. Mater. Trans. A* **1999**, *30*, 2429–2437. [CrossRef]

27. Rhodes, C.G.; Mahoney, M.W.; Bingel, W.H.; Spurling, R.A.; Bampton, C.C. Effects of friction stir welding on microstructure of 7075 aluminum. *Scr. Mater.* **1997**, *36*, 69–75. [CrossRef]

28. Ma, K.; Wen, H.; Hu, T.; Topping, T.D.; Isheim, D.; Seidman, D.N.; Lavernia, E.J.; Schoenung, J.M. Mechanical behavior and strengthening mechanisms in ultrafine grain precipitation-strengthened aluminum alloy. *Acta Mater.* **2014**, *62*, 141–155. [CrossRef]

metals

MDPI

Article

Influence of Alloys Position, Rolling and Welding Directions on Properties of AA2024/AA7050 Dissimilar Butt Weld Obtained by Friction Stir Welding

Alessandro Barbini *, Jan Carstensen and Jorge F. dos Santos

Helmholtz-Zentrum Geesthacht, Institute of Materials Research, Materials Mechanics,
Solid-State Joining Processes (WMP), 21502 Geesthacht, Germany; jan.carstensen@hzg.de (J.C.);
jorge.dos.santos@hzg.de (J.F.d.S.)
* Correspondence: alessandro.barbini@hzg.de; Tel.: +49-(0)4152-872-054; Fax: +49-(0)4152-872-033

Received: 26 February 2018; Accepted: 19 March 2018; Published: 22 March 2018

Abstract: Friction stir welding (FSW) was carried out for the butt joining of dissimilar AA2024-T3 and AA7050-T7651 aluminium alloys with 2-mm thicknesses. A comparison between the position and orientation of different materials was performed by varying the welding speed while keeping the rotational speed constant. Through an analysis of the force and torque produced during welding and a simple analytical model, the results indicate that the heat input was reduced when the AA7050 alloy was located in the advancing side (AS) of the joint. The different material positions influenced the material transportation and the interface in the centre of the stir zone (SZ). The microhardness of both materials was lower when they were in the AS of the joint. The differences in the hardness values were reduced at higher welding speeds when the heat input was decreased. The mechanical performance increased when the lower strength alloy was located in the AS. The material orientation exhibited a small influence when the AA7050 alloy was in the AS and in general on the resulting microhardness for all the cases analysed. The tensile strength values were very similar for both orientations, but an increase in the yield strength could be measured when the materials were oriented in the transverse direction.

Keywords: friction stir welding; dissimilar welded joints; materials position; material orientation; process analysis; microstructure analysis; mechanical behaviour

---

1. Introduction

In aeronautical structures, the joining throw riveting or adhesive bonding of dissimilar materials is a common practice, which is necessary to increase the mechanical performance of machines [1,2]. The problems associated with these types of joining techniques, which are well known and widely used, are related mainly to the increase in weight, the high costs of the assembly and spare parts and the difficulty in the development of an efficient automatic process for the installation.

Friction stir welding (FSW), since its invention in the 1991 at The Welding Institute (TWI Ltd., Cambridge, UK) [3], has been seen as a competitive technique for joining dissimilar materials with a good consistency in mechanical properties at a high productivity. Currently, reviews were written regarding FSW with an overview regarding process development, the influence of different tool geometries, the generated microstructure and the resulting mechanical properties [4,5].

Dissimilar welds of AA6061 and AA7075 obtained by FSW have been previously studied [6], focusing on the processing parameters and the position of the materials. Dissimilar welding involving an aluminium 6XXX series alloy was investigated previously by Amancio-Filho et al. [7] without analysing the effects of the material position on the microstructure and the mechanical properties of the welds. Another study involving more traditional alloys for the aeronautical industry has been performed by Khodir and Shibayanagi [8] with AA2024 and AA7075.

In this study, the welding speed (WS) was relatively low with a maximum speed of 3.3 mm/s, which seems further from the possibility of an industrial application. Additionally, the same alloys were investigated in terms of material flow to achieve a better understanding of the mixing process [9]. Barbini et al. investigated the combination of materials presented here without considering the influence of the material positioning, and the study was focused on a comparison between FSW and a new variant called stationary shoulder FSW [10].

Studies regarding the behavior of butt-joints of AA7050 in different heat treatment conditions obtained by FSW were performed in the past [11–13]. These analyses focused on the influence of the weld on the precipitation evolution of the different areas developed during the process and the resulting mechanical properties. Similar analyses were performed in the case of AA2024 in T3 and T351 conditions [14–16]. The influence of process parameters on the temperature profiles and the mechanical properties were investigated as well for this alloy [17]. The important role played by the rotational speed in order to favor an adequate material flow and achieve sound joints without defects was highlighted.

The general conclusion of these works is that, due to the heat generated during the process and the temperature profile experienced, for both materials, precipitate transformation was observed and coarsening in the different zones generated after welding that influenced both local and global mechanical properties of the joints.

In the present study, the influence of the process parameters and the material position of dissimilar AA2024-T3 and AA7050-T7651 were studied in terms of the microstructure and mechanical properties. The influence of the process parameters was investigated by varying the transverse speed between 3 and 8 mm/s and maintaining a constant rotational speed of 600 rpm. A variation in the welding speed influenced the heat input of the process, according to the torque-based model proposed by Khandkar et al. [18]. Two separate analyses were performed to understand the influence of the variation in the material position and orientation. The two materials were firstly switched between the advancing and retreating side of the joint to understand the effect of different compositions on the joint properties, and then the orientation of the materials and the rolling direction with respect to the welding direction was changed (Figure 1).

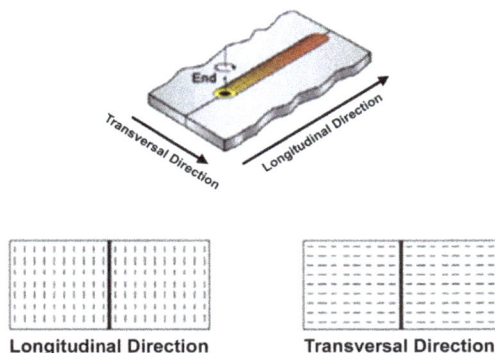

Longitudinal Direction　　　**Transversal Direction**

Figure 1. Relation between the rolling and the welding directions.

The aim of this study was to obtain a solid base of data that could be used in future studies regarding the dissimilar welds of these types of aluminium alloys and explain how all the variables involved in the process can influence the mechanical properties of the joints.

2. Experimental Procedure

2.1. Base Materials and Welding Conditions

The two base materials, AA2024-T3 and AA7050-T7651, were cut in sheets of 300×150 mm with a thickness equal to 2 mm. The typical chemical composition of the two alloys was reviewed by Dursun et al. in the study regarding the most common materials used in the aerospace industry [19]. The aluminium 2024-T3 alloy is one of the most used materials in fuselage construction due to its high strength and excellent fatigue resistance. The mechanical properties of this alloy were measured in order to compare them with the results obtained after welding and they can be found in Table 1.

Table 1. Mechanical properties of the aluminium alloy AA2024-T3.

Mechanical Properties of AA 2024-T3							
Hardness, Vickers	Yield Strength (MPa)			Tensile Strength (UTS) (MPa)			Elongation at Break (%)
	L	T	D	L	T	D	
134	379	319	325	487	474	468	20.15

AA7050 is usually produced in thick plates or extrudes and is used in the aircraft industry for the realisation of fuselage stringers and wing panels. The mechanical properties of AA7050-T7651 were also measured and are listed in Table 2.

Table 2. Mechanical properties of the aluminium alloy AA7050-T7651.

Mechanical properties of AA 7050-T7651			
Hardness, Vickers	Yield Strength (MPa)	UTS (MPa)	Elongation at Break (%)
171	490	552	11

The welds were performed in the longitudinal direction with an FSW gantry machine. The tool used to perform the welds was formed by a flat shoulder of 13 mm of external diameter with a spiral profile and a conical probe with a diameter of 5 mm at the base and 3 mm at the top (Figure 2). The probe had a left-handed thread with three flat surfaces at the sides (Triflat). The material used for the entire tool is a high-performance molybdenum-vanadium alloyed hot-work tool steel (Hotvar). The welds were realised with force control, i.e., increasing the axial force while increasing the welding speed to obtain defect free joints.

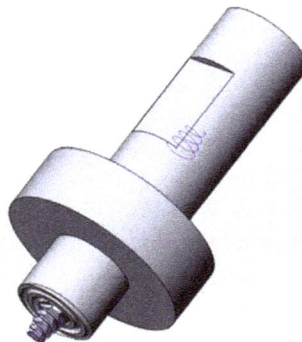

Figure 2. 3D model of the FSW tool.

In the experiment design, a rotational speed (ω) of 600 rpm and a tilt angle of $0°$ were kept constant and equal. The only process parameter that was changed was the welding speed, which modified the ratio between the rotational and welding speeds, commonly known as the pitch ratio, for each weld. The force parallel to the welding direction, called the welding force, and the torque measured during the process were recorded and used for the analysis of the process. The parameters involved in this research and the identification code of the welds are displayed in Table 3.

Table 3. Specimens' ID and welding parameters.

Specimen ID	Welding Speed (mm/s)	Material in the Advancing Side (AS)	Direction of the Materials with Respect to the Welding Direction
FSW-WS = 3-AA2024-R	3	AA2024-T3	Rolling
FSW-WS = 3-AA7050-R	3	AA7050-T7651	Rolling
FSW-WS = 3-AA2024-P	3	AA2024-T3	Perpendicular
FSW-WS = 3-AA7050-P	3	AA7050-T7651	Perpendicular
FSW-WS = 5-AA2024-R	5	AA2024-T3	Rolling
FSW-WS = 5-AA7050-R	5	AA7050-T7651	Rolling
FSW-WS = 5-AA2024-P	5	AA2024-T3	Perpendicular
FSW-WS = 5-AA7050-P	5	AA7050-T7651	Perpendicular
FSW-WS = 8-AA2024-R	8	AA2024-T3	Rolling
FSW-WS = 8-AA7050-R	8	AA7050-T7651	Rolling
FSW-WS = 8-AA2024-P	8	AA2024-T3	Perpendicular
FSW-WS = 8-AA7050-P	8	AA7050-T7651	Perpendicular

2.2. Characterisation Methods

The samples necessary to analyse the microstructure and hardness were cut orthogonally to the welding line with a length of 50 mm. For an investigation of the joints' cross section, the samples were etched with Dix–Keller Reagent with an immersion time of 15 s after undergoing standard metallographic preparation.

The microhardness tests were performed on a Zwick/Roell ZHV machine (Ulm, Germany) with an applied load of 0.2 kgf for 10 s, in accordance with the standard ASTM E384-10. The line of indentation was positioned at the middle of the sample thickness, namely, 1 mm from the upper surface. The distance between the indentations was 0.5 mm, and the total length of the horizontal profile was 30 mm, symmetrical with respect to the weld centre (Figure 3).

Figure 3. Microhardness measurement indentations across a FSW joint.

A three-point bending test was performed to certify the quality of the joint immediately after the welding process. According to the standard ASTM E190-92, the diameter of the mandrel was 8 mm, and the distance between the supports was 15.2 mm. The specimens were tested in a Zwick/Roell universal testing machine with a load capacity of 100 kN. The test was stopped when the drop in the load reached 75% of the maximum force applied, and, in this way, all the specimens were tested under the same conditions. After the end of the test, the angle reached by the specimens was measured.

The tensile tests were performed orthogonally to the welding direction on the standard specimens with a 12.5 × 2 mm cross section on a Zwick/Roell universal testing machine with a load capacity of 100 kN. The strain was measured using a mechanical extensometer (MTS Systems GmbH, Berlin, Germany) with a gage length of 50 mm, positioned with its centre in the weld nugget. The tests were executed following the standard ASTM E8-09 at a room temperature of 22 °C with a constant transverse speed of 1 mm/min. For each welding condition, three specimens were used.

It is important to remark that, before proceeding with the mechanical tests, a period of more than 40 days was permitted to allow the process of natural ageing of the materials [11] and to ensure stable mechanical properties.

3. Results and Discussion

In the current study, all the characterisation methods and analyses were performed in a way to distinguish the effect of a different positioning of the materials from a different orientation. For this reason, firstly, the influence of the position of the material with respect to the weld line was analysed, and, subsequently, the influence of the orientation of the material was also analysed.

3.1. Position of the Materials

The first step was to investigate the influence of the materials position on the measured forces during the process and the consequent heat generation. After that, the resulting microstructure and material properties developed were analyzed in depth to have a clear understanding of the joint performance.

3.1.1. Process Analysis

The analysis of the process was based on the influence of a different material position on the welding force and torque measured during the process. Due to the asymmetric heat generation and material transportation between the (AS) and retreating side (RS) [20], it was expected that the difference in strength between the two materials would affect the two measured values.

When considering the welding force (Figure 4), note that, at the two lower welding speeds, the force on the tool is lower when the high-strength material is in the AS. The higher heat transferred in the AS causes a larger plasticisation of AA7050, reducing the global resistance of the material on the tool. This behaviour ceases at a higher welding speed, where the reduced heat transferred into the weld does possibly lead to an increase in the resistance of AA7050 to be transported around the probe. When this material is located in the RS, it was not transported but just extruded by the probe, causing a smaller influence of the heat on the resisting force.

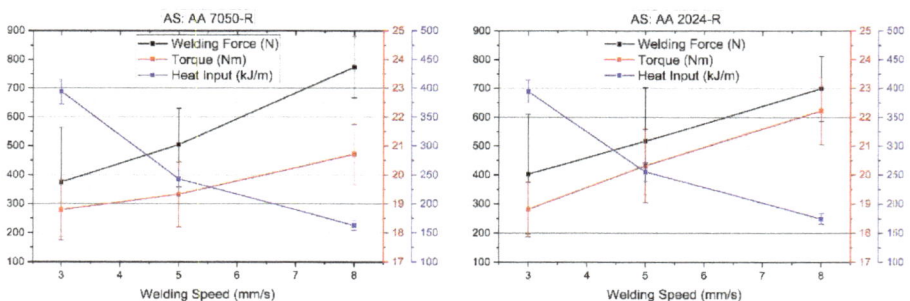

Figure 4. Measured forces and calculated energy input for different material positions.

In the case of the torque, a similar consideration was accomplished with some differentiators. The torque at the lower WS is the same for both material positions. With the AA2024 alloy in the AS, the trend of an increase in torque with the welding force is linear with a slope higher than what was initially measured for the other material positioning configuration. For this reason, at medium and high welding speeds, the torque measured for AA2024 in the AS is higher than that for the other variant. This higher torque could be connected to the lower strength of AA2024 that leads to a larger amount of material to be transported around the probe and consequently to an increase in the torque. The measured torque in AA7050 in the AS is not linear, but, similar to the welding force, increased sharply at the higher WS. Once again, this effect could be related to a different condition of the AA7050 alloy when the temperature decreased, which led to larger measured forces.

The rate of heat generation was calculated according to the formula used to characterise the energy transferred into the weld through the analysis of the torque measured during the process [21]. The simplified version of this equation is shown in Equation (1):

$$Q = \frac{2\pi\omega T}{v},\tag{1}$$

where Q is the energy input per unit length, T the measured torque and v is the welding speed.

The trend of the heat input shown in Figure 4 is similar for both material positions. A steeper reduction could be calculated between the low and medium welding speeds with a curve that tends to flatten at a higher WS. When the calculated values are compared, a slight reduction in the heat generated could be measured in AA7050 in the AS.

3.1.2. Microstructural Analysis

For a clearer understanding of the analysis of all the generated microstructures, the AS of the joint was always located on the left-hand side of the pictures.

In Figure 5, the microstructures of the joints obtained at 3 mm/s, with in the AS AA7050 (Figure 5a) and AA2024 (Figure 5b), are shown.

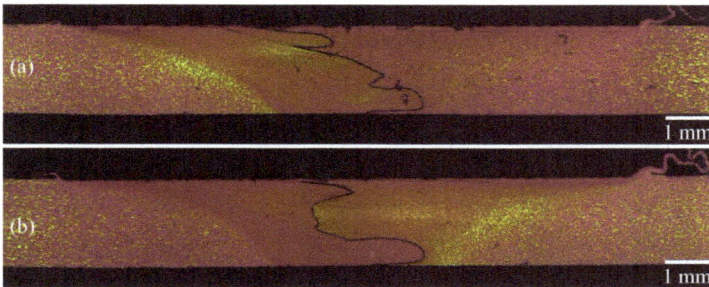

Figure 5. Cross sections of (**a**) FSW-WS = 3-AA7050-R and (**b**) FSW-WS = 3-AA2024-R.

Note that, in the AS, the border between the Stir Zone (SZ) and the Thermo-Mechanically Affected Zone (TMAZ) was more defined for both configurations. This difference is due to the asymmetry of the process, which considers both material transport and heat transfer. The material in the AS was sheared around the probe and the shoulder, leading to higher heat generation by the strain energy in this location than in the RS. In the RS, the material around the probe underwent a smaller rotation and was extruded outward [22,23]. In the case of AA2024-T3 positioned in the RS (Figure 5a), it was almost impossible to distinguish between the SZ and the TMAZ at this level of magnification. This implies a more homogenous transition from the SZ to the TMAZ, but, more importantly, the dynamic recrystallisation that generates a grain size reduction in the SZ is less marked [13]. This finding could be explained by the higher strength of the AA7050-T7651 alloy, when positioned in the AS, which increases the resistance on the probe reducing the flow of the material and the effect of the shear forces on the RS in the last part of the probe rotation.

Another noticeable feature that was observed from the images was the difference in the mixture of the material in the SZ. Due to the nature of these dissimilar joints, the material flow in the centre was observed after etching through the presence of the so-called "onion rings," the concentric rings in the SZ marked in the figure. In the case of AA2024-T3 in the AS (Figure 5b), this particular structure, appearing after joint formation, is almost symmetric with respect to the longitudinal axis. The symmetry of the material mixture could be seen as a positive aspect, which implies a better material transportation through all the thickness that led to a more homogeneous structure of the weld. On the other hand, the behaviour of AA7050 in the AS was anti-symmetric, as shown in Figure 5a, which exhibits the influence of alloy strength on material transportation.

Similar observations also apply in the case of increased welding speed for the macrographs shown in Figure 6.

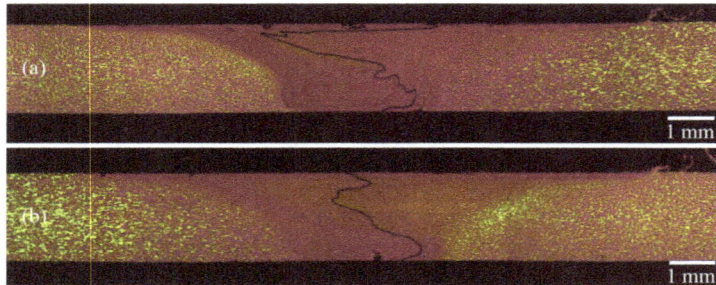

Figure 6. Cross sections of (**a**) FSW-WS = 5-AA7050-R and (**b**) FSW-WS = 5-AA2024-R.

For AA7050-T7651 in the AS (Figure 6a), when increasing the welding speed, the material homogeneity was improved since the higher WS caused a decrease in the heat input, which was more consistently distributed through the sample thickness. When AA2024 was placed in the AS (Figure 6b), the interface of the material started to lose the symmetric behaviour that was previously shown, with an increase in the black line tilting. In the case of AA7050 in the AS, the WS demonstrated less influence on the material mixing showing similar characteristics to what was already observed. The reduction in the SZ on the root side of the weld shows that, at higher welding speeds, the amount of material stirred by the shoulder was less influenced by the WS than the one stirred by the probe. For this reason, the interface tends to increase the tilting (Figure 6b).

With an additional increase in the welding speed, note a decrease in the Heat Affected Zone (HAZ) for both the configurations showed in Figure 7. The HAZ in AA2024 was still wider when this material was positioned in the RS.

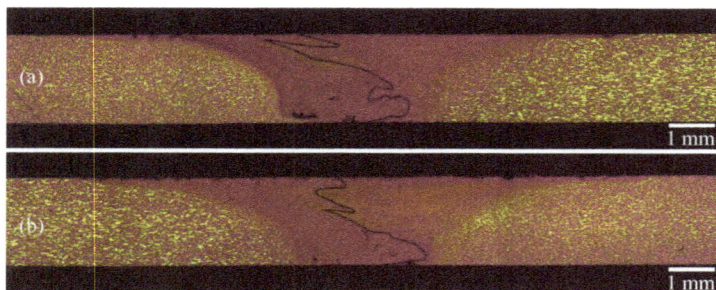

Figure 7. Cross sections of (**a**) FSW-WS = 8-AA7050-R and (**b**) FSW-WS = 8-AA2024-R.

The shape of the onion rings was very similar in both cases, showing that, by increasing the WS, the two materials interface was less influenced by the strength of the alloy positioned in the AS or RS of the joint. However, the difference in material transportation between the two materials positions could be seen from the generation of a tunnel defect at the bottom of the SZ in the AS of the joint (Figure 8).

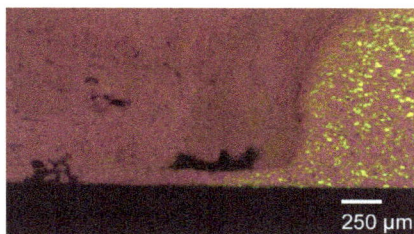

Figure 8. Defect detail generated in the specimen FSW-WS = 8-AA7050-R.

The defect is located in the area where the weld temperature decreases and also the stirring mechanism of the probe is reduced due to the tapered geometry and the consequent reduction of the tangential velocity. This defect was found only in the case when the AA7050 was placed in the AS of the joint and consequently underwent a larger straining that could not be followed due to the reduced ductility of the base material (see Table 2).

3.1.3. Microhardness Analysis

To directly compare the hardness trend in the two materials, AA2024 was always positioned on the left side of the following graphs. Consequently, the AS and RS of the joints are inverted in the two hardness profiles plotted in each figure. Figure 9 showed the hardness profile for a WS = 3 mm/s. In this figure, the diameters of the shoulder and the probe are represented to correlate the variation in the hardness profile with the tool used for the process.

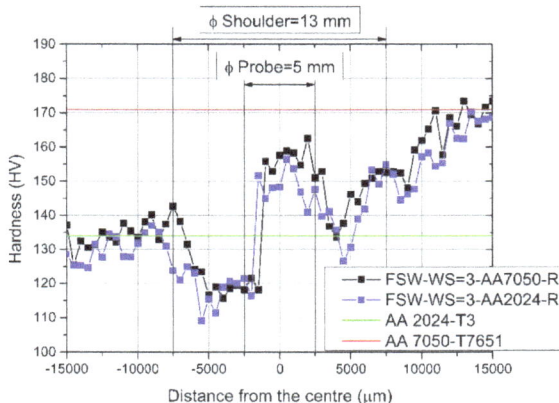

Figure 9. Microhardness test results for WS = 3 mm/s and the different positioning of the materials.

In both of the cases represented, it was possible to see an increase in the hardness in the proximity of the probe where the SZ of the weld was located. This increment was due to the combined actions of the shear forces applied by the probe on the material and the dynamic recrystallisation that caused grain refinement in the SZ. However, the main factor influencing the hardness in this area, where the highest temperature is reached, is correlated with the dissolution of precipitates and the possibility for the solutes to precipitate and age again [24].

In the space between the probe and the shoulder, the hardness dropped significantly, especially in the AS. In this part, it is possible to identify two different zones in the weld, namely, the TMAZ and the HAZ. These areas are characterised by the absence of direct actions of the probe on the material, while the heat generated by the weld was still affecting them. The shear layers of the plasticised material in direct contact with the probe and the shoulder influenced the material in the TMAZ. In this area, the hardness reached the minimum in both sides of the joint and for the two different positions of the materials. The temperature reached in the TMAZ is not high enough to favor the dissolution of precipitates that would instead transform and coarsen, reducing drastically the hardness and the strength of the material. In the AA2024, the Guinier-Preston-Bagaryatsky (GPB) zones are gradually transforming into S'(S) strengthening precipitates that would increase in percentage with the temperature and subsequently coarsen. The precipitates in the AA7050 undergo a similar transformation evolving from super solute to Guinier-Preston (GP) zones and in the TMAZ become coarsened $\eta(MgZn_2)$ precipitates that show a minimum hardness.

On the other hand, in the HAZ, there is no mechanical action introduced in the material, and the modification of the microstructure is only due to the transmission of heat. The hardness increased gradually in the AS and RS until the original value of the respective base material was attained. Only small differences in the hardness profile between the two material positions were evident. Moreover, it was possible to observe that both the materials reached their original hardness more rapidly when they were positioned in the AS. This consideration was important especially for the AA2024 side, where failure was more probable.

When increasing the WS (Figure 10), it is possible to notice a divergence in the hardness behaviour at the side of the AA7050-T7651 alloy. When this material was positioned in the AS, the hardness increased faster due to the sharper transition between the SZ and HAZ previously seen in the microstructural observations.

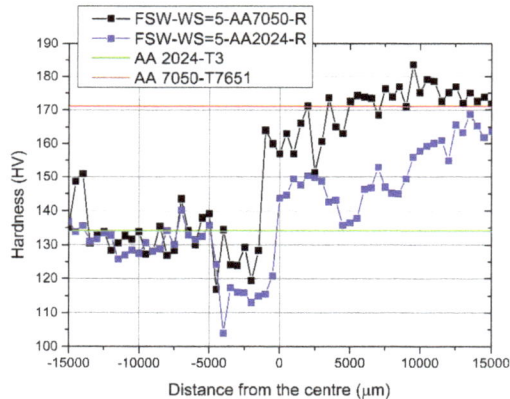

Figure 10. Microhardness test results for WS = 5 mm/s and the different positioning of the materials.

In the weaker part of the weld, there was no significant dissimilarity in the TMAZ from the side of AA2024-T3. The difference in the two cases between the minimum values reached by the hardness profiles is 5 HV that, according to Tabor's equation [25], should lead to a difference in the yield stress of approximately 18 MPa.

In Figure 11, the case when the WS increased until 8 mm/s was shown. The hardness performance was improved for both the situations, especially in the case of the AA2024-T3 in the AS. Here, the minimum of the hardness was greater than 120 HV, and the variation was limited to a small area close to the probe diameter.

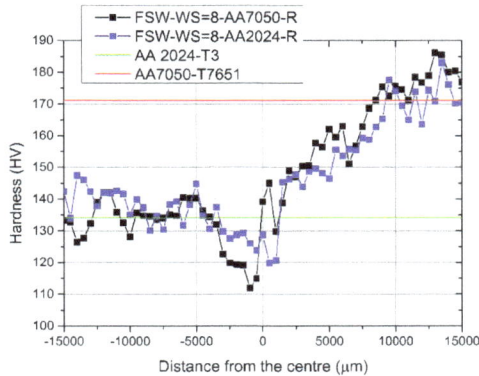

Figure 11. Microhardness test results for WS = 8 mm/s and the different positioning of the materials.

An increase in the WS and consequently a decrease in the heat input leads to a lower temperature peak during the weld and causes a smoother hardness profile. All of the typical welding zones were reduced in size, especially in the proximity of the weld centre. Once again, both of the materials returned to their original hardness values more rapidly when they were in the AS, even if the difference between the two cases was minimal for the highest WS.

3.1.4. Mechanical Characterisation

In the bending analysis, the minimum angle that the specimen should reach to pass the test was 80°, considering that the base material testing of AA7050-T7651 broke at an angle of approximately 90°. Table 4 shows the results obtained from changing the position of the materials.

Table 4. Results of the bending test for the material position.

Specimen ID	Bending Test Results	
	Begin of the Weld	End of the Weld
FSW-WS = 3-AA2024-R	V	V
FSW-WS = 3-AA7050-R	V	X
FSW-WS = 5-AA2024-R	V	V
FSW-WS = 5-AA7050-R	X	X
FSW-WS = 8-AA2024-R	V	V
FSW-WS = 8-AA7050-R	X	X

All of the specimens except one, AA7050-T7651 in the AS and in the rolling direction, did not pass the test. This result showed a preliminary influence of the material location on the mechanical properties of the joints, considering that the entire set of specimens with the material in the rolling direction and AA2024-T3 have to overcome this step of 80°.

Figure 12a illustrates the variation in the yield stress as a function of increasing WS for AA 2024-T3 in both the advancing side and the retreating side.

For both of the cases, the yield stress increased with increasing welding speed. This finding agrees with the microhardness test results and the macrograph investigation. The reduction in the heat input caused a shrinking in the TMAZ and HAZ sizes with smaller grain growth. The peak of the temperature during the process was reduced, which influenced the precipitate transformation and coarsening. Furthermore, when the WS increased, the forging force of the probe increased, which applied larger shear forces to the material that generated a more refined grain structure with better mechanical properties.

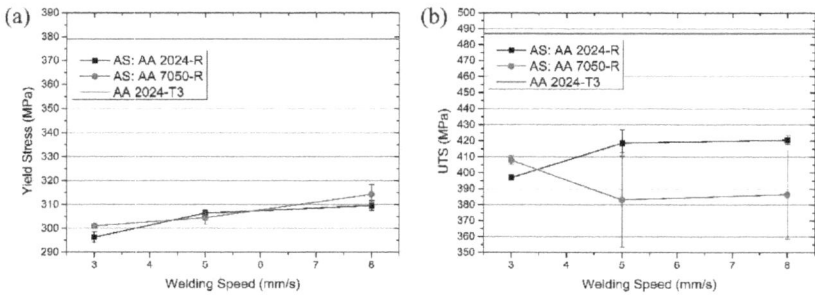

Figure 12. Variation in the (**a**) Yield Stress (YS) and (**b**) UTS with changing WS and the material position.

The values of the yield stress, considering also the scatter, did not show significant variations in the changing position of the material. The maximum difference obtained from the test is 5 MPa. The difference between the positions of the two materials was represented by an enlargement in the scattering of the results when AA7050-T7651 was positioned in the AS and the welding speed was increased. The reason for this behaviour could be caused by a major sensitivity to external factors, such as the non-perfect clamping of the plates, or internal factors such as a variation in the precipitate distribution and residual stresses in the material that caused a less stable process.

This last observation was confirmed by the joint strength analysis presented in Figure 12b. When increasing the WS, the UTS dropped drastically, and the scatter in the results increased to high levels for AA7050 in the AS of the weld. This finding confirmed the instability of the process along the welds. The analysis of the fracture location for the different welding conditions showed that, for the low welding speed and for all the specimens with AA2024 in the AS, the fracture happened in the TMAZ where the minimum hardness was measured (Table 5). This is clearly explained by the degradation mechanism of the precipitates due to the high temperatures. At higher welding speeds with the AA7050 in the AS, i.e., when the tensile strength dropped, the failure location moved toward the centre of the SZ. In this case, the reason for the premature failure was dependent on the unstable plasticisation when the temperature decreased that, combined with the higher strength and lower ductility of AA7050, did not allow the complete mixing and interlocking of the materials in the centre of the joint.

Table 5. Fracture location with changing WS and the material position.

Specimen ID	Fracture Location
FSW-WS = 3-AA2024-R	AS-TMAZ
FSW-WS = 3-AA7050-R	RS-TMAZ
FSW-WS = 5-AA2024-R	AS-TMAZ
FSW-WS = 5-AA7050-R	SZ
FSW-WS = 8-AA2024-R	AS-TMAZ
FSW-WS = 8-AA7050-R	SZ

When AA2024 was in the AS, a WS of 8 mm/s and a UTS value of 86.4% of the base material was obtained. The scatter, at all WSs, remained under 3% of the absolute value measured, confirming the stable behaviour of the weld along all the line.

The materials position clearly influenced the strain mechanism around the tool and the heat generation. When the AA7050 was located in the AS of the weld, the transportation of the material is reduced as certified by the decreased torque and heat input due to the higher strength and lower ductility of the alloy. While an improvement in the precipitates evolution could be attested for in this configuration for the two lower welding speeds, as seen from the hardness profile measured, this did not bring any benefit in the joint performance under tensile and bending tests. The ductility of the joint was reduced and, at higher welding speeds, the fracture location moved to the interface between the two alloys in the SZ centre. The problem of positioning the AA7050 in the AS is that it resulted as impossible to improve joint performances due to the reduced strength at low welding speeds caused by the coarsening of precipitates in the TMAZ, where the AA2024 is located, and the lack of material transportation at higher welding speeds that led to a reduction of the bonding mechanism between the two alloys and the formation of defects. On the other hand, when the AA2024 was placed in the AS, the joint properties could be improved by increasing the welding speed, hence reducing the heat input in the weld and the area affected by it, without excessively reducing the material transportation around the tool that would cause a transition of the fracture to the SZ.

3.2. Direction of the Materials

The influence of the rolling orientation in respect to the welding direction on the joints properties was performed in a similar way to what was previously done for the materials' positioning. The investigation starts with the analysis of the influence on the heat generation to subsequently see how this affects the microstructure and mechanical characteristics of the joints.

3.2.1. Process Analysis

For the analysis on the process, the measured values for AA2024 in the rolling direction and in the AS were used as a reference. The material position and orientation were changed to show both the alloys in the AS and perpendicular to rolling orientation (Figure 13).

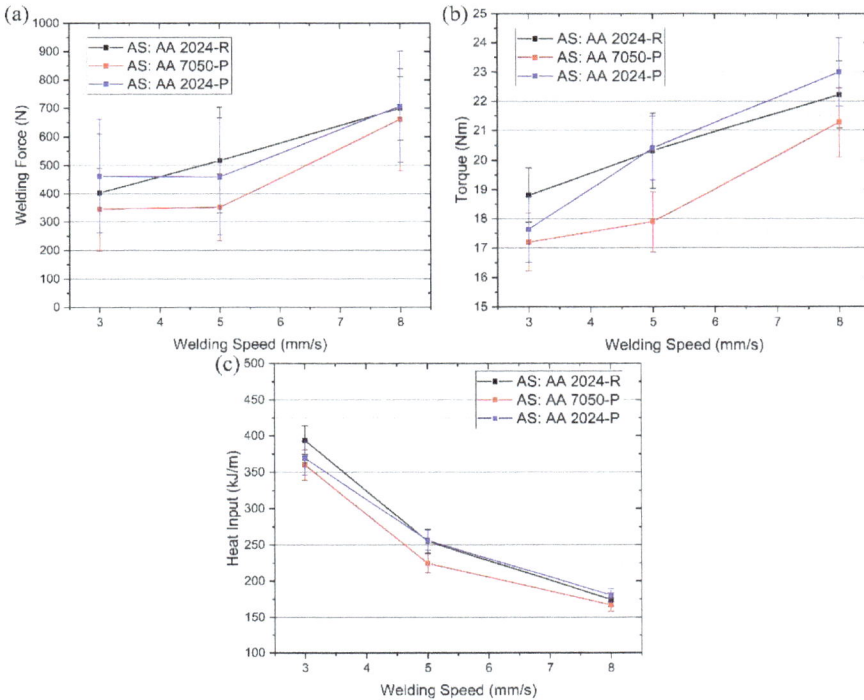

Figure 13. Measured: (**a**) Welding force, (**b**) Torque and (**c**) calculated energy input for the different material orientations.

In a comparison of the curves for AA2024-T3 in the AS, all of the parameters analysed were similar and in the limit of the tolerance when the welding speed reached 5 mm/s. The influence of the material orientation was evident at the lower WSs where the contact between the tool and the surrounding material in each rotation increased. The anisotropy of the base material in the two main directions led to an increase in the force (Figure 13a) due to the lower deformability in the transverse direction of the material in front of the tool and a decrease in the torque (Figure 13b) due to the lower strength. The results are also similar when comparing AA7050-T7651 in the AS for the two orientations.

3.2.2. Microstructural Analysis

Figure 14 represents the case of AA2024 in the AS of the weld for the two directions of the materials and a low WS.

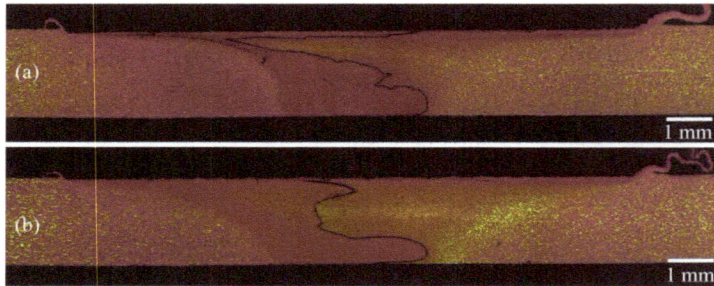

Figure 14. Cross sections of (**a**) FSW-WS = 3-AA2024-P and (**b**) FSW-WS = 3-AA2024-R.

In the case of a material orientation perpendicular to the rolling direction, the zone affected by the shoulder was more limited, and the shape of the SZ was parallel to the probe at the root side of the weld. With the material disposed in the rolling direction, the slope of the borders between the SZ and TMAZ is less steep. The two HAZs on both the sides of the SZ have comparable dimensions with no noticeable difference in the shape. The mixing of the materials was influenced by their orientation, and a more asymmetric interface was exhibited for the sample shown in Figure 14a.

In AA7050 in the AS, there were no remarkable differences between the two macrographs (Figure 15). The sizes of the all the characteristic welding zones were not influenced significantly by the orientation of the materials.

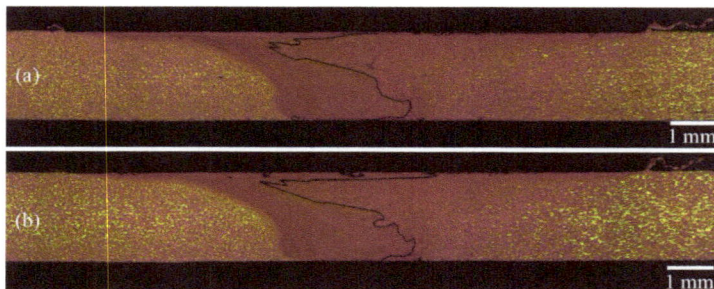

Figure 15. Cross section of (**a**) FSW-WS = 5-AA7050-P and (**b**) FSW-WS = 5-AA7050-R.

The "onion rings" did not change shape or orientation, indicating that the rolling or perpendicular to the rolling directions are not influenced by material transportation. In general, it appears that, when a high-strength material is in the AS of the joint, both the welding parameters and the material orientation demonstrate a lower influence on this interface.

3.2.3. Microhardness Analysis

In Figure 16a, the influence of the direction of the materials can be seen in the case of the average WS and for the AA2024-T3 alloy positioned in the AS of the weld.

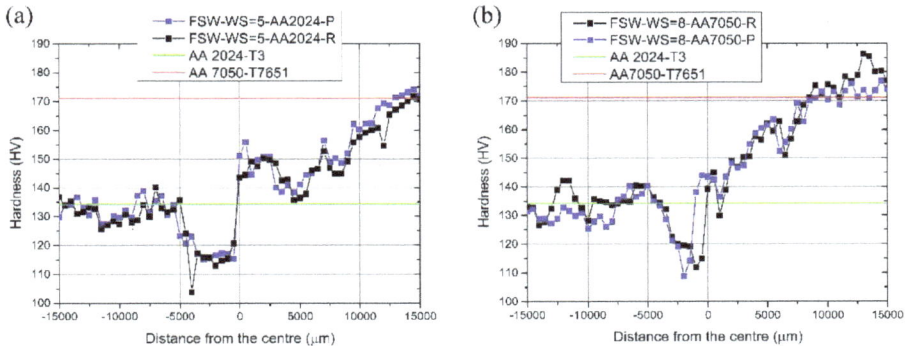

Figure 16. Microhardness for different material orientations: (**a**) WS = 5 mm/s, AA2024AS and (**b**) WS = 8 mm/s, AA7050AS.

No remarkable differences can be seen in the hardness profile with a change in the direction of the materials. The lowest point in the case of the plates in the rolling direction is probably due to the presence of a lower hardness precipitate in the indentation area. The same conclusion could be drawn for the highest WS and with AA7050-T7651 in the AS (Figure 16b). The only difference is a slight movement of the curve in the proximity of the SZ to the AA7050 side for the plates disposed in the rolling direction.

From this study, is it possible to conclude that the direction of the materials does not influence the hardness profile for different WSs and material dispositions, i.e., its values or shape. The low difference previously shown in the calculated heat input does not justify variations in the precipitation mechanism for different material orientations; hence, no variation in the hardness profile could be expected.

3.2.4. Mechanical Characterisation

The bending test results are listed in Table 6. The results of the test obtained for the AA2024-T3 alloy in the AS and in rolling direction were used as a comparison.

Table 6. Results of the bending test for the material direction and AA2024-T3 in the AS.

Specimen ID	Bending Test Results	
	Begin of the Weld	End of the Weld
FSW-WS = 3-AA2024-R	V	V
FSW-WS = 3-AA2024-P	V	X
FSW-WS = 5-AA2024-R	V	V
FSW-WS = 5-AA2024-P	V	X
FSW-WS = 8-AA2024-R	V	V
FSW-WS = 8-AA2024-P	X	X

None of the specimens in the direction perpendicular to the rolling direction passed the test when welded with the highest WS.

The analysis of the yield stress was performed in a similar manner to that performed during the process analysis with AA2024-T3 in the AS and in the rolling direction used as reference for the other two cases (Figure 17a). The same procedure was followed in the study of the joint strength.

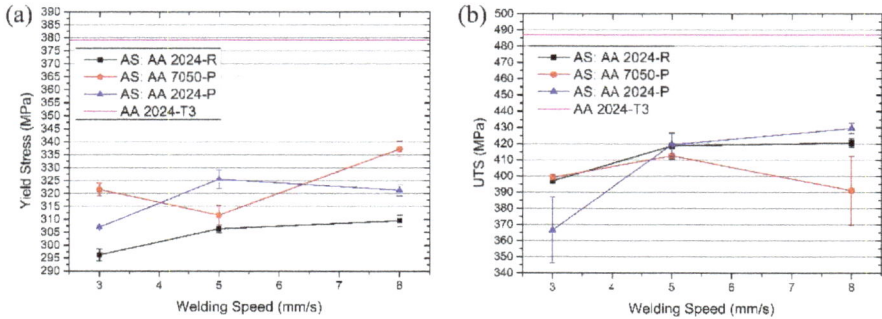

Figure 17. Variation in the (**a**) YS and (**b**) UTS with changing WS, the material position and direction.

The yield stress obtained in plates disposed perpendicularly to the rolling direction is higher for all the WSs than that in the rolling direction. In the case of the plates in the rolling direction, the yield stress improved with increasing WS, while, in the other two cases, there is a variability that does not allow any assumption regarding possible trends in the process. The difference in the yield stress with the changing direction of the materials is significant, especially when comparing the values at higher welding speeds. The explanation is related to the different strengths of the base materials in the two main directions and the lower influence of the modified microstructures at lower load levels.

For the ultimate tensile test (Figure 17b), the behaviour of the weld with AA7050 in the AS was not influenced by the orientation of the materials. The mechanical properties of the joint decrease when the WS increases and their stability is poor. The fracture behaviour was also similar to the parent configuration with the materials in the rolling direction in Table 7. The final failure was mostly located in the centre of the SZ at the interphase between the two alloys.

Table 7. Fracture location with changing WS, the material position and direction.

Specimen ID	Fracture Location
FSW-WS = 3-AA2024-R	AS-TMAZ
FSW-WS = 3-AA2024-P	SZ
FSW-WS = 3-AA7050-P	SZ
FSW-WS = 5-AA2024-R	AS-TMAZ
FSW-WS = 5-AA2024-P	AS-TMAZ
FSW-WS = 5-AA7050-P	RS-TMAZ
FSW-WS = 8-AA2024-R	AS-TMAZ
FSW-WS = 8-AA2024-P	AS-TMAZ
FSW-WS = 8-AA7050-P	SZ

In the case of AA2024-T3 in the AS, similar results were obtained at middle and high WSs for both material directions. The best result for the high WS reached 88.2% of the base material with an increment of the 2% with respect to the configuration in the rolling direction. At low WS, the tensile strength, when the orientation of the materials was perpendicular to the rolling direction, was considerably lower with a high scatter of the values. Once again, in this case, the fracture location moved to the centre of the SZ at the interphase. As previously shown (Figure 14), the alloys interphase in the case of orientation perpendicular to the rolling direction resulted in being asymmetric and similar to the one obtained for AA7050 in the AS. This specific configuration is probably leading to a reduction in joining force in the centre of the weld due to poor material transportation.

4. Conclusions

The influence of the direction and position of the materials on the microstructure and mechanical performance of the welds was systematically analysed. An analysis of the process showed a reduction in the heat generated with the AA7050-T7651 material positioned in the AS of the weld caused by a general decrease in the torque. The material orientation influences the heat generated with a smaller reduction when the two alloys are oriented perpendicular to the rolling direction. This small variation is due to the anisotropy of the rolled base materials.

There is a relevant advantage in positioning the AA2024-T3 alloy in the AS of the weld, which was verified by all the tests performed, especially considering the results obtained from the bending test and the tensile test. In this last test, even with a decrease in the yield stress, the improvement in the ultimate tensile stress is remarkable and shows the best support for this conclusion. This increase was justified by the better material transportation that led to a failure in the TMAZ where the minimum hardness was measured. Another important point is the higher stability of the tensile tests outcome for AA2024-T3 in the AS of the weld, confirmed by the lower standard deviation of the results for all the WSs considered.

The material orientation is a more complicated matter, and the results do not show a clear direction to take when choosing between the rolling or perpendicular direction in order to improve the joint strength. On one hand, the best results in the bending test were obtained by positioning the plates in the rolling direction since all the specimens passed the test. Meanwhile, the tensile test showed better behaviour regarding the yield stress in the case of the material disposition perpendicular to the rolling direction, while only a small difference between the two configurations was present considering the ultimate tensile stress. At a lower WS, the weld realised with the plates in the rolling direction exhibited a higher ultimate stress and a reduced scatter of the measured values. At a microstructural point of view, the differences were minimal at the magnitude of the macrograph used. The hardness profiles for the two cases analysed were almost identical, which suggests a slight influence of the direction of the material in the evolution of the precipitates.

Acknowledgments: This work embedded in the Research Platform ACE has been carried out within the scope of the Research Platform "Light-weight Assessment, Computing and Engineering Centre" (ACE Centre) as part of the reference project Lightweight Integral Structures for Future Generation Aircrafts (LISA).

Author Contributions: A.B. and J.F.d.S. conceived the design of experiments; A.B. performed the experiments analysed the data and wrote the article; J.C. performed the welds and contributed to the process analysis.

Conflicts of Interest: The authors declare no conflict of interest.

References

1. Li, G.; Shi, G.; Bellinger, N.C. 6—Assessing the riveting process and the quality of riveted joints in aerospace and other applications A2—Chaturvedi, M.C. In *Welding and Joining of Aerospace Materials*; Woodhead Publishing: Sawston, UK, 2012; pp. 181–214.
2. Kwakernaak, A.; Hofstede, J.; Poulis, J.; Benedictus, R. 8—Improvements in bonding metals for aerospace and other applications A2—Chaturvedi, M.C. In *Welding and Joining of Aerospace Materials*; Woodhead Publishing: Sawston, UK, 2012; pp. 235–287.
3. Thomas, W.M.; Nicholas, E.D.; Needham, J.C.; Murch, M.G.; Temple-Smith, P.; Dawes, C.J. Improvements to Friction Welding. Patent EP 065,326,5A2, 6 December 1991.
4. Mishra, R.S.; Ma, Z.Y. Friction stir welding and processing. *Mater. Sci. Eng.* **2005**, *50*, 1–78. [CrossRef]
5. Nandan, R.; DebRoy, T.; Bhadeshia, H.K.D.H. Recent advances in friction-stir welding—Process, weldment structure and properties. *Prog. Mater. Sci.* **2008**, *53*, 980–1023. [CrossRef]
6. Guo, J.F.; Chen, H.C.; Sun, C.N.; Bi, G.; Sun, Z.; Wei, J. Friction stir welding of dissimilar materials between AA6061 and AA7075 al alloys effects of process parameters. *Mater. Des.* **2014**, *56*, 185–192. [CrossRef]

7. Amancio-Filho, S.T.; Sheikhi, S.; dos Santos, J.F.; Bolfarini, C. Preliminary study on the microstructure and mechanical properties of dissimilar friction stir welds in aircraft aluminium alloys 2024-T351 and 6056-T4. *J. Mater. Process. Technol.* **2008**, *206*, 132–142. [CrossRef]
8. Khodir, S.A.; Shibayanagi, T. Friction stir welding of dissimilar AA2024 and AA7075 aluminum alloys. *Mater. Sci. Eng. B* **2008**, *148*, 82–87. [CrossRef]
9. Da Silva, A.A.M.; Arruti, E.; Janeiro, G.; Aldanondo, E.; Alvarez, P.; Echeverria, A. Material flow and mechanical behaviour of dissimilar AA2024-T3 and AA7075-T6 aluminium alloys friction stir welds. *Mater. Des.* **2011**, *32*, 2021–2027. [CrossRef]
10. Barbini, A.; Carstensen, J.; dos Santos, J.F. Influence of a non-rotating shoulder on heat generation, microstructure and mechanical properties of dissimilar AA2024/AA7050 FSW joints. *J. Mater. Sci. Technol.* **2018**, *34*, 119–127. [CrossRef]
11. Fuller, C.B.; Mahoney, M.W.; Calabrese, M.; Micona, L. Evolution of microstructure and mechanical properties in naturally aged 7050 and 7075 Al friction stir welds. *Mater. Sci. Eng. A* **2010**, *527*, 2233–2240. [CrossRef]
12. Zhou, L.; Wang, T.; Zhou, W.L.; Li, Z.Y.; Huang, Y.X.; Feng, J.C. Microstructural characteristics and mechanical properties of 7050-T7451 aluminum alloy friction stir-welded joints. *J. Mater. Eng. Perform.* **2016**, *25*, 2542–2550. [CrossRef]
13. Su, J.Q.; Nelson, T.W.; Mishra, R.; Mahoney, M. Microstructural investigation of friction stir welded 7050-T651 aluminium. *Acta Mater.* **2003**, *51*, 713–729. [CrossRef]
14. Genevois, C.; Fabrègue, D.; Deschamps, A.; Poole, W.J. On the coupling between precipitation and plastic deformation in relation with friction stir welding of AA2024 T3 aluminium alloy. *Mater. Sci. Eng. A* **2006**, *441*, 39–48. [CrossRef]
15. Sutton, M.A.; Yang, B.; Reynolds, A.P.; Taylor, R. Microstructural studies of friction stir welds in 2024-T3 aluminum. *Mater. Sci. Eng. A* **2002**, *323*, 160–166. [CrossRef]
16. Bousquet, E.; Poulon-Quintin, A.; Puiggali, M.; Devos, O.; Touzet, M. Relationship between microstructure, microhardness and corrosion sensitivity of an AA 2024-T3 friction stir welded joint. *Corros. Sci.* **2011**, *53*, 3026–3034. [CrossRef]
17. Carlone, P.; Palazzo, G.S. Influence of process parameters on microstructure and mechanical properties in AA2024-T3 friction stir welding. *Metallogr. Microstruct. Anal.* **2013**, *2*, 213–222. [CrossRef]
18. Khandkar, M.Z.H.; Khan, J.A.; Reynolds, A.P. Prediction of temperature distribution and thermal history during friction stir welding: Input torque based model. *Sci. Technol. Weld. Join.* **2003**, *8*, 165–174. [CrossRef]
19. Dursun, T.; Soutis, C. Recent developments in advanced aircraft aluminium alloys. *Mater. Des.* **2014**, *56*, 862–871. [CrossRef]
20. Tongne, A.; Desrayaud, C.; Jahazi, M.; Feulvarch, E. On material flow in friction stir welded Al alloys. *J. Mater. Process. Technol.* **2017**, *239*, 284–296. [CrossRef]
21. Cui, S.; Chen, Z.W.; Robson, J.D. A model relating tool torque and its associated power and specific energy to rotation and forward speeds during friction stir welding/processing. *Int. J. Mach. Tools Manuf.* **2010**, *50*, 1023–1030. [CrossRef]
22. Chen, Z.W.; Pasang, T.; Qi, Y. Shear flow and formation of nugget zone during friction stir welding of aluminium alloy 5083-O. *Mater. Sci. Eng. A* **2008**, *474*, 312–316. [CrossRef]
23. Arbegast, W.J. A flow-partitioned deformation zone model for defect formation during friction stir welding. *Scr. Mater.* **2008**, *58*, 372–376. [CrossRef]
24. Zhang, Z.; Xiao, B.L.; Ma, Z.Y. Hardness recovery mechanism in the heat-affected zone during long-term natural aging and its influence on the mechanical properties and fracture behavior of friction stir welded 2024Al–T351 joints. *Acta Mater.* **2014**, *73*, 227–239. [CrossRef]
25. Tabor, D. The physical meaning of indentation and scratch hardness. *Br. J. Appl. Phys.* **1956**, *7*, 159. [CrossRef]

metals

MDPI

Article

A Correlation between the Ultimate Shear Stress and the Thickness Affected by Intermetallic Compounds in Friction Stir Welding of Dissimilar Aluminum Alloy–Stainless Steel Joints

Florent Picot [1,2], Antoine Gueydan [1], Mayerling Martinez [1], Florent Moisy [1] and Eric Hug [1,*]

[1] Laboratoire de Cristallographie et Sciences de Matériaux, Normandie Université, ENSICAEN, CNRS UMR 6508, 6 Boulevard du Maréchal Juin, 14050 Caen, France; florent.picot@ensicaen.fr (F.P.); antoine.gueydan@ensicaen.fr (A.G.); mayerling.martinez@ensicaen.fr (M.M.); florent.moisy@ensicaen.fr (F.M.)
[2] Sominex, 13 rue de la Résistance, 14400 Bayeux, France
* Correspondence: eric.hug@ensicaen.fr; Tel.: +33-231-451-313

Received: 19 February 2018; Accepted: 9 March 2018; Published: 13 March 2018

Abstract: In this work, Friction Stir Welding (FSW) was applied to join a stainless steel 316L and an aluminum alloy 5083. Ranges of rotation and translation speeds of the tool were used to obtain welding samples with different heat input coefficients. Depending on the process parameters, the heat generated by FSW creates thin layers of Al-rich InterMetallic Compound (IMC) mainly composed of $FeAl_3$, identified by energy dispersive spectrometry. Traces of Fe_2Al_5 were also depicted in some samples by X-ray diffraction analysis and transmission electron microscopy. Monotonous tensile tests performed on the weld joint show the existence of a maximum mechanical resistance for a judicious choice of rotation and translation speeds. It can be linked to an affected zone of average thickness of 15 µm which encompass the presence of IMC and the chaotic mixing caused by plastic deformation in this area. A thickness of less than 15 µm is not sufficient to ensure a good mechanical resistance of the joint. For a thickness higher than 15 µm, IMC layers become more brittle and less adhesive due to high residual stresses which induces numerous cracks after cooling. This leads to a progressive decrease of the ultimate shear stress supported by the bond.

Keywords: FSW process; aluminum alloy; stainless steel; intermetallic compounds; mechanical strength

1. Introduction

Unlike traditional welding methods, Friction Stir Welding (FSW) is an assembly technique which occurs without additional metal and does not reach the melting point of the materials [1]. Numerous fields of application can find advantages of the process such as automotive and railway industries [2]. Critical technological fields such as air transport, the development of fuel tanks for aerospace applications and the nuclear industry also use FSW to join alloys [3]. The FSW technology opens the possibility of joining materials difficult to weld by traditional fusion processes, such as Mg/Steel [4], Al/Ti [5], Al/Mg [6], and Al/Cu [7] combinations. However, industrial joining between such dissimilar materials still remains a technological challenge because of the numerous parameters which could affect the joint quality.

Numerous studies concerning FSW were performed focusing on different aspects of the process: tool material [8], tool shoulder geometry [9], pin global geometry [10] and thread [11], material flow, and heat generated during the welding [12,13]. The joint has been analyzed by residual stress measurements [14] and microstructure characterization [15]. Moreover, it is well known that FSW results in the formation of layers of InterMetallic Compounds (IMC) through the interface. The covalent bonds in IMC increase the binding energy and decrease the number of available free electrons, generally increasing the brittleness of the junction [16]. However, for an optimal thickness, intermetallics provide good bonding characteristics as long as the layer remains compact. Previous studies therefore aim to determine the existence of an optimal intermetallic layer thickness as far as the mechanical properties are concerned [17].

Stainless steel 316L and aluminum alloys 5083 are often used in the transport industry but welding them together remains difficult. Some previous studies concerning this combination focused on the microstructure evolution [18,19] in different joining configurations. In butt joining configuration, some investigations dealt with the relation between mechanical strength and the existence of stainless steel particles in aluminum [20] or on IMC growth following the main process parameters [18]. Methods using the Taguchi technique extract the influence on the welding quality for each processing parameter [21]. In lap join configuration, probe penetration in the lower part influences the welding quality [22]. The mechanical strength of the joint drastically decreases if the probe does not penetrate the lower level of the lap configuration. On the probe path, the tool revolution speed and welding speed affect the grain size reduction and the mechanical strength [15]. From a metallurgical point of view, aluminum-stainless steel FSW method mainly creates $FeAl_3$ intermetallic compound which can decrease the mechanical strength of the junction [23] for a critical IMC thickness higher than 20 µm typically [24].

The main objective of this work is to present a methodology of FSW using a lap join configuration developed to weld a stainless steel 316L–aluminum alloy 5083 combination and the way to optimize the mechanical strength of the junction. To this end, the link between the mechanical strength of the weld and the IMC thickness induced by the process heat input was investigated. It was demonstrated that an optimal thickness can be reached by an adequate choice of process parameters.

2. Materials and Methods

For this study, a FSW configuration (Sominex, Bayeux, France) is used, inspired from Kimapong and Watanabe [24]. In such a configuration, a 5083 aluminum sheet (4.85 mm in thickness) covers a 316L stainless steel sample (3.5 mm in thickness). Chemical composition of the alloys is given in Table 1.

The tool in Figure 1a entirely goes through the aluminum alloy and scratches the surface of the stainless steel. The depth penetration inside the stainless steel is held at 0.35 mm. On the welding zone, the aluminum alloy sheet is entirely stirred by the pin with a 3° tilt (Figure 1b,c). The tool is made of tungsten carbide and has a 12 mm diameter flat shoulder. The probe has a threaded conical shape and is about 5 mm in length. The end of the pin has a 4 mm diameter corresponding to the width of the welded zone.

Table 1. Chemical composition (wt. %) of aluminum 5083 and stainless steel 316L alloys.

Aluminum 5083 (Al Balance)							
Mg	Si	Fe	Cu	Mn	Cr	Zn	Ti
4.0–4.9	0.40	0.40	0.10	0.25	0.05–0.25	0.20	0.15
Stainless Steel 316L (Fe Balance)							
C	Si	Mn	Cr	Ni	Mo	-	-
0.025	0.40	1.20	16.80	10.10	2.10	-	-

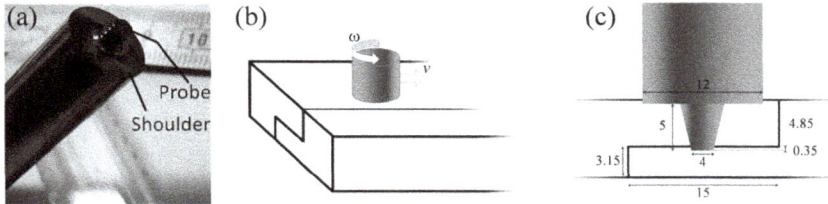

Figure 1. Friction stir welding setup. (**a**) FSW tool used for all the samples. (**b**) Schematic representation of the tool in the lap joining configuration and tool parameters. (**c**) Lap join friction stir welding cross section (dimensions in mm).

During the welding, local warming of the zone can take place, mainly generated by friction between metal and shoulder. The temperature locally reaches 0.6 to 0.75 times the melting point of aluminum alloy [1]. In addition, the rotational movement of the tool produces a flow of material from the front to the back of the pin which plastically deforms and compresses the material around the shoulder. This phenomenon induces important microstructural changes [25]. Two process parameters were studied [26]: the rotational speed ω of the tool, ranging from 600 rpm to 2100 rpm, and the welding speed v in the 10–100 mm·min^{-1} range. The two parameters were studied from a Taguchi ANOVA DoE (Design of Experiment) plan with a L$_{16}$ resolution in order to localize the optimal region of the process parameters [27]. A model based on dimensional analysis, developed by Roy et al. [28], was used to estimate a non-dimensional heat input Q^* during welding, which is expressed by Equation (1). Since the coefficient of friction changes with temperature, it is difficult to accurately calculate the corresponding heat generation. This parameter strongly depends on the ratio f between thermal properties of the materials at the tool/aluminum alloy interface (Equation (2)).

$$Q^* = \frac{f\sigma_{Y80}A\omega C_P}{kv^2},$$ (1)

with:

$$f = \left(\frac{k^{Al}\rho^{Al}C_P^{Al}}{k^T\rho^T C_P^T}\right)^{1/2} = 0.971,$$ (2)

k represents the thermal conductivity, ρ is the density and C_P is the specific heat. σ_{Y80} is the yield stress of the upper material (aluminum alloy) at 80% of the solidus temperature and $A = 10^{-4}$ m^2 is the cross-section area of the tool. The combination between the translation speed v of the tool and its rotational velocity ω enables to calculate Q^* which ranges from 0.20 to 64 in this study. Input constant values are listed in Table 2.

Table 2. Input constants of aluminum [29] and WC tool [30] for Q^* computation.

Aluminum				Tool		
k^{Al} (W·m^{-1}·K^{-1})	ρ^{Al} (g·cm^{-3})	C_P^{Al} (J·K^{-1}·kg^{-1})	σ_{Y80} (MPa)	k^T (W·m^{-1}·K^{-1})	ρ^T (g·cm^{-3})	C_P^T (J·K^{-1}·kg^{-1})
117	2.66	900	7.5	110	13.30	203

Tensile samples were cut perpendicular to the welding direction as in Figure 2 in order to obtain the ultimate shear strength (USS) τ_{max} supported by the weld. Samples were tested on a 5569 dual column machine (INSTRON, Norwood, MA, USA) with a constant displacement of 1 mm·min^{-1}. Monotonous tensile tests were performed using traditional ASTM standards on 20 mm junction length samples from a 200 mm length of plates. The junction was positioned in the middle of the 55 mm in the tensile machine jaws space. For each Q^* value, at least three tensile tests were performed in order to take into account the dispersion of the results. The welded junction is parallel to the tensile direction and the surface S which supports the shear strength is the welding bead length (about 20 mm) of the sample by the probe width (4 mm). τ_{max} is obtained by the ratio between the maximal strength F_{max} and the surface S.

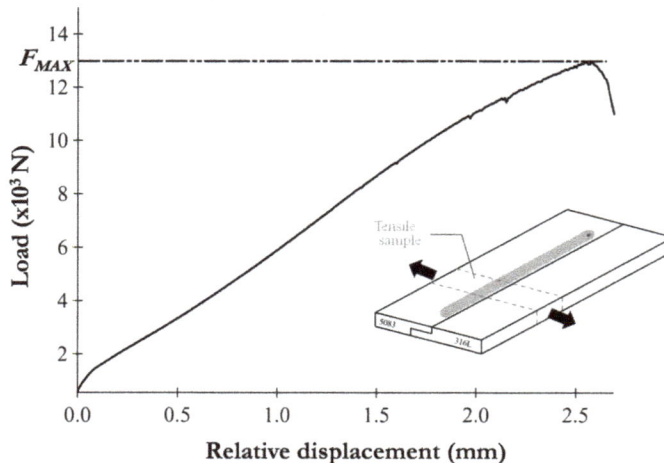

Figure 2. Typical mechanical test curve ($Q^* = 48$). In inset: Sample cutting pattern on a weld bead.

Microscopic observations of the interface were carried out by Scanning Electron Microscopy (SEM ZEISS SUPRA 55 EDS, Marly le Roi, France). The analysis of composition was performed by Energy Dispersive Spectrometry (EDS) on mechanical polished specimens. Complementary phase identification was carried out using X-ray diffractometer (Bruker, Conventry, UK) with monochromatic Cu K_α radiation and continuous scan mode at $0.25°/$min over a wide angle range of 30–90°. Transmission Electron Microscopy (TEM) (JEOL 2010, Croissy-sur-Seine, France) operating at 200 kV enables microstructural observations with dark field imaging and IMC identification with diffraction patterns and additional local EDS measurements. Thin foils for TEM observations were prepared in cross sectional configuration at the interface by Focused Ion Beam (FIB) method. This was performed with a dual beam FEI 660 (Nanolab Inc., Waltham, MA, USA) operating with Ga+ ions at 30 keV.

3. Experimental Results

Figure 3 shows typical cross sections perpendicular to the welding path of the junction for increasing values of Q^*. The path of the pin is clearly visible at the interface between the stainless steel (light grey) and the aluminum alloy (dark grey). For some values of Q^*, two long intrusions of stainless steel can be observed around the part, inside the aluminum part (Figure 3c). These metal shapes are not necessarily symmetric and were already mentioned between two aluminum sheets as Cold Lap Defects (CLD) [31]. The biggest CLD is mostly at the retreating side and generates an additional strength which contributes to the resistance of the weld. When the depth penetration of the tool increases, the CLDs are larger. However, for lower values of Q^*, as it can be seen in Figure 3b, these defects appear to be unstable, crushed in small parts and dispersed in the aluminum alloy.

In addition, microstructural observations of the interface also exhibit the presence of ultra-fine intermetallic compounds in the form of discontinuous thin shapes in stainless steel, especially visible in Figure 3a,d. This can be attributed to the presence of intermetallic compounds as it was already mentioned by Kimapong and Watanabe [24] for A5083 aluminum–SS400 steel couple. When the heat coefficient Q^* is higher than 45, the IMCs grow faster and interface appears as a multilayer composed of stainless steel 316L matrix and rich-(FeAl) IMCs (Figure 4). The thermal expansion coefficient between stainless steel ($\alpha_{ss} = 16$–18 $\mu m \cdot m^{-1} \cdot K^{-1}$), aluminum ($\alpha_{Al} = 23.4$ $\mu m \cdot m^{-1} \cdot K^{-1}$) and IMC

(α_{FeAl3} = 14 µm·m^{-1}·K^{-1}) [32] vary in a wide range and create high residual stresses which induce cracks all along the interface after the cooling of the samples. Depending on Q^*, the intermetallic compound grows first perpendicular to the interface and either continues in the same direction, or changes direction to grow parallel to the 5083/316L interface. This results are in good agreement with previous observations [33].

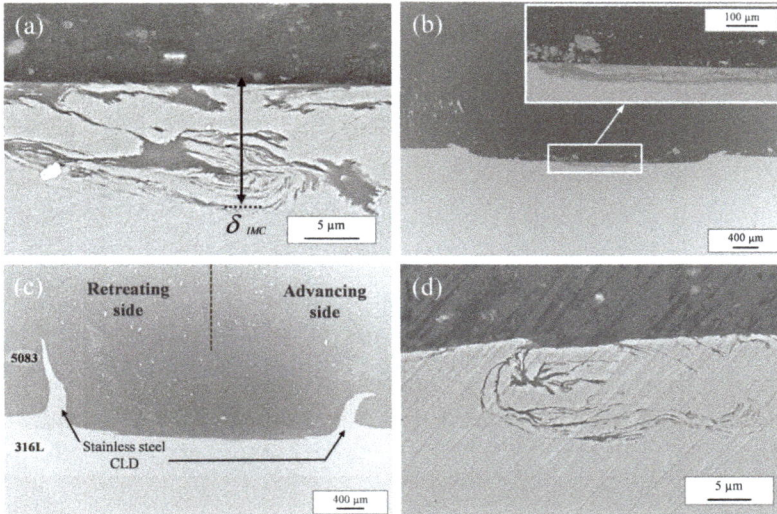

Figure 3. Cross section and InterMetallic Compound (IMC) protrusion at the welded interface for increasing values of Q^*. (**a**) Examples of measurement of the characteristic depth of IMC δ_{IMC}. (Q^* = 0.39). (**b**) Stainless steel fragmentation around the welded interface at low heat input (Q^* = 2.5). (**c**) Cross section observation of welded interface and Cold Lap Defects (CLD) highlights (Q^* = 56.4). (**d**) IMC infiltration profile at higher Q^* values (Q^* = 64.1).

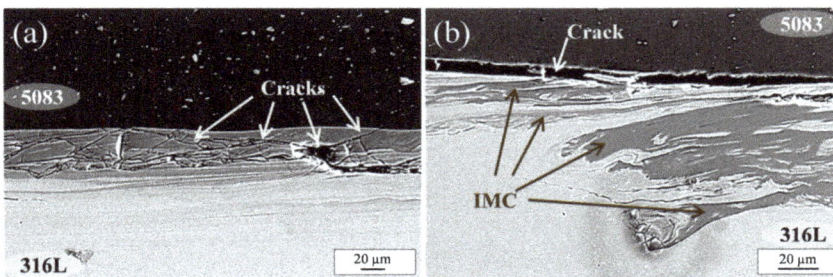

Figure 4. Scanning Electron Microscopy (SEM) interface observation for high values of Q^*. (**a**) Broken IMC at the interface Q^* = 48.01, (**b**) crack along the interface Q^* = 56.4.

In order to identify the nature of the IMC, EDS was performed on several zones rich in intermetallics. Chemical analyses reveal the presence of an Al-rich IMC of FeAl$_3$ nominal composition as evidenced in Figure 5a,b. However, the binary phase diagram Fe/Al contains five different intermetallic compounds [34]. Others studies by Girard et al. [35] and Nishida et al. [23] pointed out that not all intermetallics predicted by equilibrium phase diagram are present after stir welding process, the principal IMC in both studies being FeAl$_3$ (Fe$_4$Al$_{13}$). This could be explained by the fact that FSW is a complex process of plastic deformation and heating, far from the equilibrium thermodynamic conditions.

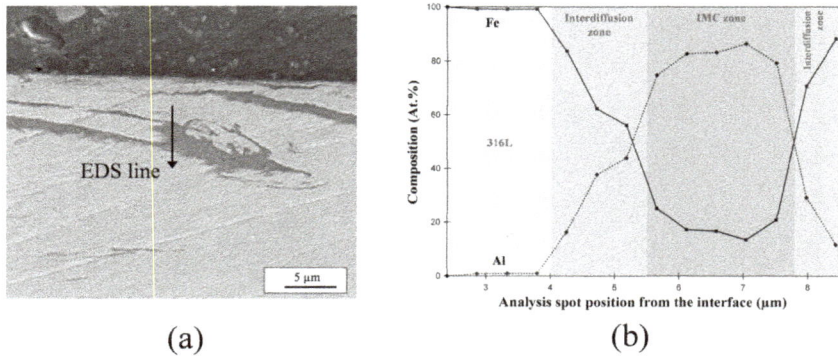

(a)   (b)

Figure 5. Identification of the intermetallic compound by Energy Dispersive Spectrometry (EDS). (a) EDS measurement line through the IMC zone at the welded interface (Q^* = 0.86). (b) Atomic composition along the EDS line.

XRD measurements (Figure 6) confirm the existence of $FeAl_3$ but also exhibit traces of Fe_2Al_5. This second IMC was not identified for all process conditions and can be related to specific values of Q^*. Deeper insight into the IMC layer at the interface is provided by TEM. The principal results are included in Figure 7. The bottom of the figure represents a reconstructed view of the thin lamella. The initial materials can be readily identified. The aluminum alloy (in the right part of the picture) is composed of grains of uniform size (around 3 μm), one of the representative diffraction pattern of the cubic structure of aluminum is shown (grey arrow). The aluminum side contains also iron rich particles. Such particles could pre-exist in the parent Al material but also are probably pieces of steels torn off during the welding process that were afterwards enriched in aluminum by diffusion.

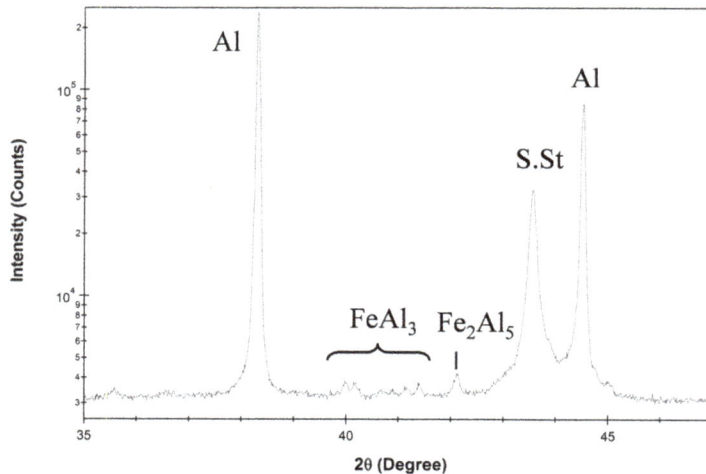

Figure 6. XRD spectra analysis focusing on typical Fe–Al IMC peaks in the range of 35–47° (S.St. is the abbreviation of Stainless Steel).

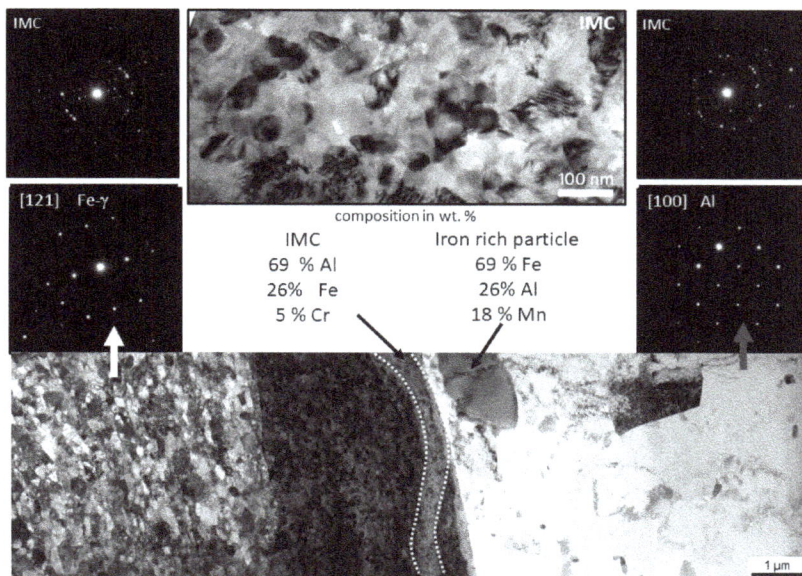

Figure 7. Identification of the intermetallic compounds by Transmission Electron Microscopy (TEM) showing the presence of traces of Fe_2Al_5 and $FeAl_2$.

The stainless steel, at the left part of the image, exhibits submicron-sized grains, the size of which decreases when approaching the interface. The electron diffraction analysis was performed at around 6 μm from the interface where grains can be easily selected for identification (white arrow). It turns clear that the process affects the initial microstructure of steel producing nanostructured grains near the interface in the 0.35 mm thick strip of material.

Another clearly identifiable zone is highlighted with dotted lines along the interface. This zone of a wavy shape can be associated with the IMC. The microstructure of IMC is shown in the top image of Figure 7 together with representative selected area electron diffraction (SAED) patterns. The identification of this nanostructured intermetallic was not possible. Diffraction patterns were in all tested cases a complex mixture of diffraction of grains with different orientations.

EDS analysis of the IMC zone seems to confirm that there is the presence of the $FeAl_3$ and/or Fe_2Al_5 compounds. However, due to their relatively close stoichiometric ratios, the nanometer size of grains and to the fact that the analysis is rather semi-quantitative, it is difficult to distinguish reliably the two compounds.

The highly disturbed character of the interface makes it difficult to analyze the different phases. In addition, it is known that $FeAl_3$ first grows and then decreases due to the formation of Fe_2Al_5 upon traditional welding [36]. Finally, it is also noticeable that the different compositions highlighted by EDS could reveal the existence of a more complex rich-$FeAl_3$ IMC, enriched in chromium, originating from the initial composition of the stainless steel. Indeed, FSW process enables a fast atomic diffusion through the interface allowing the nucleation and growth of intermediate phases.

All of these results allow concluding that upon the FSW process there are many phenomena that occur (mechanical and thermal) at the interface, which allow the formation of highly mixed zone, comprising of different IMC compounds such as $FeAl_3$ and Fe_2Al_5.

In order to quantify the mechanical resistance of the weld, USS values of the most relevant tests are given in Table 3 in function of Q^*.

Table 3. Average value of the maximum shear strength τ_{max} for various values of Q^*.

Q^*	0.21	0.39	0.86	2.52	3.37	48.01	23.94	56.36	64.11
τ_{max} (MPa)	95 ± 20	127 ± 12	106 ± 25	130 ± 12	92 ± 7	83 ± 3	70 ± 4	62 ± 15	124 ± 4

The heat process strongly influences the mechanical resistance of the junction for the Stainless Steel–Al alloy combination in contrast to others combinations as for instance the Al-Ti system [37]. Even for the lowest value of Q^* ($Q^* = 0.21$), the joint exhibits a broad range of shear strength values between 85–105 MPa. τ_{max} sensitively increases with Q^* until reaching an optimum around 130 MPa for $Q^* = 2.52$ (the error bars also decrease here). For higher values of Q^*, τ_{max} decreases down to 62 MPa for $Q^* = 56$. However, for greatest values of Q^* (close to 64), τ_{max} exhibits again high levels, around 125 MPa. This phenomenon is closely linked to the interface organization of intermetallic compounds, coupled with the existence of mechanical anchoring created from the tool path. The numerous cracks depicted along the interface for Q^* higher than 60 imply therefore that the welds are not acceptable from an industrial point of view.

4. Discussion—Correlation between Ultimate Shear Strength, Heat Input, and IMC Depth

This part discusses in more details the different depths of the IMC zone, as a function of the different values of Q^*, and their influence on the mechanical resistance of the junction. Even if it is difficult to precisely identify the nature and a precise thickness of IMC phases at the interface of the junction, an IMC depth (δ_{IMC}), corresponding to the thickness affected by the presence of these IMCs can be estimated. An example showing an estimation of δ_{IMC} is given in Figure 3a. These δ_{IMC} values were measured for each value of Q^*, and represented in Figure 8a. Three distinct stages can be identified in this figure. In the first stage (low heat input Q^* belongs to [0.2–2.5]), the thickness of stainless steel affected by IMCs remains almost constant with a δ_{IMC} value around 9 μm. This stage is representative of the nucleation of rich-FeAl$_3$ IMCs followed by its growth which appear relatively slow and controllable. This is the combination of the mechanical and thermal phenomenon from the welding process which allows IMCs emergence even at very low provided heat input. After a critical value $Q^* = 2.5$, the second stage is related to a strong increase in δ_{IMC}. During this stage, rich-FeAl$_3$ extends and widens inside the volume of stainless steel leading to IMC percolation mechanisms. In the third stage, related to high values of Q^* ($Q^* > 48$), the IMC grows less in depth in the stainless steel and becomes more parallel to the interface, which implies a decrease of δ_{IMC}. This last step represents a stage of overgrowth for rich-FeAl$_3$ IMCs. When higher values of Q^* are reached (typically $Q^* = 64$), $\delta_{IMC} \approx 10$ μm which corresponds to the same order as values reached in the first stage. However, the shape of IMC layers is clearly different. In this stage, δ_{IMC} corresponds to a single layer originated from interdiffusion mechanisms. This can be theoretically described with the traditional Matano-Boltzmann approach for a two-phase system [38,39].

The mechanical resistance of the weld is represented by the USS value τ_{max} obtained by shear lap tests. τ_{max} is displayed as a function of δ_{IMC} in Figure 8b. It can be observed, an increase of τ_{max} until an optimum value of around 130 MPa for $\delta_{IMC} = 15$ μm. When δ_{IMC} is higher than 15 μm, τ_{max} strongly decreases. It is noticeable that, as explained in the previous paragraph, for relatively close values of δ_{IMC}, very different shapes of IMC layers can be depicted. From these results, it is thus clear that the ultimate shear stress of the weld can be related mainly to the thickness of the stainless steel affected by the rich-FeAl$_3$ IMCs, as shown in Figure 8b independently of their shapes. This curve highlights the existence of an optimal $\delta_{IMC} \approx 15$ μm giving the maximum mechanical resistance of the interface. For higher values of δ_{IMC}, τ_{max} decreases drastically and reaches a lower asymptote about 80 MPa for $\delta_{IMC} \geq 50$ μm. The increase in δ_{IMC} leads to non-cohesive and brittle IMC layers and the maximal shear strength therefore decreases.

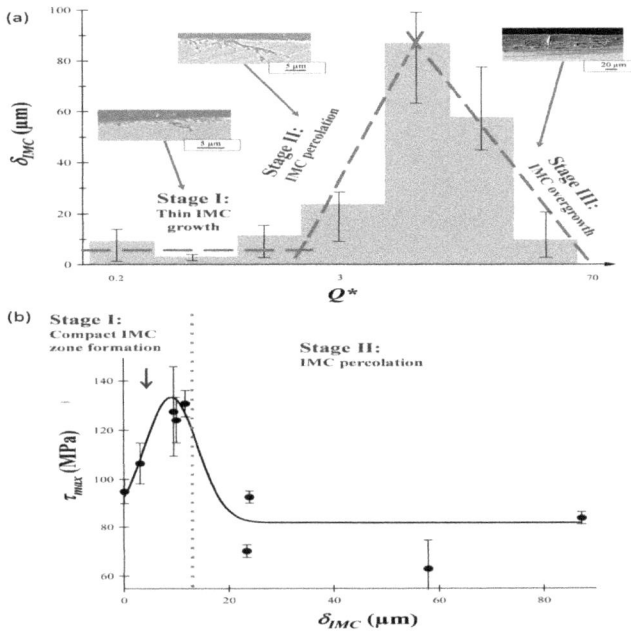

Figure 8. Correlation between τ_{max}, Q^* and the IMC affected zone. (a) δ_{IMC} plotted against Q^* (logarithmic scale for Q^*). (b) Maximum shear strength in function of δ_{IMC}.

5. Conclusions

FSW process was conducted between a stainless steel 316L and an aluminum alloy 5083 in order to determine the optimum parameters giving the higher mechanical resistance of the weld. A lap joint configuration with a lower part penetration was used with the aluminum sheet covering the 316L sample. It was possible in this work to link the heat coefficient during the welding process to the ultimate shear stress and to the thickness of stainless steel affected by the existence of rich-FeAl$_3$ IMC. Main results of this research can be summarized as follows:

- Rich-FeAl$_3$ compound was the only intermetallic detected along the weld interface independent of the FSW parameters.
- IMC growth mechanisms are linked to the heat input coefficient Q^* with three distinct stages of formation which influence the mechanical behavior of the interface.
- For an optimal thickness of about 15 µm inside the stainless steel, rich-FeAl$_3$ IMCs are present under the form of thin compact layers ensuring a good chemical cohesion of the weld. High values of the mechanical resistance of the weld are reported (130 MPa of shear strength as optimum value) in these optimal conditions.
- For higher values of the IMC thickness, ultimate shear stress dramatically decreases up to a minimal value of 80 MPa on average, and numerous cracks are depicted along the interface. The cohesion of the weld is only ensured by the existence of cold lap defects which act as mechanical anchors between the two samples.

Author Contributions: The paper is the result of a collaboration of all co-authors, Florent Picot performed this work during his Ph.D. Thesis, under the supervision of Eric Hug, Ph.D. director. Antoine Gueydan contributes to the study of experimental results, Mayerling Martinez and Florent Moisy performed TEM and XRD experiments.

Conflicts of Interest: The authors declare no conflict of interest.

References

1. Thomas, W.M.; Threadgill, P.L.; Nicholas, E.D. Feasibility of friction stir welding steel. *Sci. Technol. Weld. Join.* **1999**, *4*, 365–372. [CrossRef]

2. Thomas, W.M.; Nicholas, E.D. Friction stir welding for the transportation industries. *Mater. Des.* **1997**, *18*, 269–273. [CrossRef]

3. Dey, H.C.; Ashfaq, M.; Bhaduri, A.K.; Rao, K.P. Joining of titanium to 304l stainless steel by friction welding. *J. Mater. Process. Technol.* **2009**, *209*, 5862–5870. [CrossRef]

4. Kasai, H.; Morisada, Y.; Fujii, H. Dissimilar fsw of immiscible materials: Steel/magnesium. *Mater. Sci. Eng. A* **2015**, *624*, 250–255. [CrossRef]

5. Chen, Y.; Liu, C.; Liu, G. Study on the joining of titanium and aluminum dissimilar alloys by friction stir welding. *Open Mater. Sci. J.* **2011**, *5*, 256–261. [CrossRef]

6. Pourahmad, P.; Abbasi, M. Materials flow and phase transformation in friction stir welding of al 6013/mg. *Trans. Nonferrous Met. Soc. China* **2013**, *23*, 1253–1261. [CrossRef]

7. Xue, P.; Xiao, B.L.; Ni, D.R.; Ma, Z.Y. Enhanced mechanical properties of friction stir welded dissimilar Al–Cu joint by intermetallic compounds. *Mater. Sci. Eng. A* **2010**, *527*, 5723–5727. [CrossRef]

8. Rai, R.D.; Bhadeshia, H.K.D.H.; DebRoy, T. Friction stir welding tools. *Sci. Technol. Weld. Join.* **2011**, *16*, 325–342. [CrossRef]

9. Li, D.; Yang, X.; Cui, L.; He, F.; Zhang, X. Investigation of stationary shoulder friction stir welding of aluminum alloy 7075-t651. *J. Mater. Process. Technol.* **2015**, *219*, 112–123. [CrossRef]

10. Mishra, R.S.; Ma, Z.Y. Friction stir welding and processing. *Mater. Sci. Eng. R* **2005**, *50*, 1–78. [CrossRef]

11. Lin, Y.C.; Chen, J.N. Influence of process parameters on friction stir spot welded aluminum joints by various threaded tools. *J. Mater. Process. Technol.* **2015**, *225*, 347–356. [CrossRef]

12. Mohanty, H.K.; Venkateswarlu, D.; Mahapatra, M.M.; Kumar, P.; Mandal, N.R. Modeling the effects of tool probe geometries and process parameters on friction stirred aluminium welds. *J. Mech. Eng. Autom.* **2012**, *2*, 74–79. [CrossRef]

13. Reynolds, A.P. Visualisation of material flow in autogenous friction stir welds. *Sci. Technol. Weld. Join.* **2000**, *5*, 120–124. [CrossRef]

14. Zapata, J.; Toro, M.; Lopez, D. Residual stresses in friction stir dissimilar welding of aluminum alloys. *J. Mater. Process. Technol.* **2016**, *229*, 121–127. [CrossRef]

15. Bisadi, H.; Tour, M.; Tavakoli, A. The influence of process parameters on microstructure and mechanical properties of friction stir welded al 5083 alloy lap joint. *Am. J. Mater. Sci.* **2011**, *1*, 93–97. [CrossRef]

16. Hug, E.; Bellido, N. Brittleness study of intermetallic (Cu,Al) layers in copper clad aluminium thin wires. *Mater. Sci. Eng. A* **2011**, *A528*, 7103–7106. [CrossRef]

17. Kimapong, K.; Watanabe, T. Effect of welding process parameters on mechanical property of fsw lap joint between aluminum alloy and steel. *Mater. Trans.* **2005**, *46*, 2211–2217. [CrossRef]

18. Lan, S.; Liu, X.; Ni, J. Microstructural evolution during friction stir welding of dissimilar aluminum alloy to advanced high-strength steel. *Int. J. Adv. Manuf. Technol.* **2016**, *82*, 2183–2193. [CrossRef]

19. Yazdipour, A.; Heidarzadeh, A. Effect of friction stir welding on microstructure and mechanical properties of dissimilar al 5083-h321 and 316l stainless steel alloys joints. *J. Alloys Compd.* **2016**, *680*, 595–603. [CrossRef]

20. Yazdipour, A.; Heidarzadeh, A. Dissimilar butt friction stir welding of al 5083-h321 and 316l stainless steel alloys. *Int. J. Adv. Manuf. Technol.* **2016**, *87*, 3105–3112. [CrossRef]

21. Chen, T. Process parameters study on fsw joint of dissimilar metals for aluminum-steel. *J. Mater. Sci.* **2009**, *44*, 2573–2580. [CrossRef]

22. Elrefaey, A.; Gouda, M.; Takahashi, M.; Ikeuchi, K. Characterization of aluminum/steel lap joint by friction stir welding. *J. Mater. Eng. Perform.* **2005**, *14*, 10–17. [CrossRef]

23. Nishida, T.; Ogura, T.; Nishida, H.; Fujimoto, M.; Takahashi, M.; Hirose, A. Formation of interfacial microstructure in a friction stir welded lap joint between aluminium alloy and stainless steel. *Sci. Technol. Weld. Join.* **2014**, *19*, 609–616. [CrossRef]

24. Kimapong, K.; Watanabe, T. Friction stir welding of aluminum alloy to steel. *Weld. J.* **2004**, *83*, 277–282.

25. Threadgill, P.L.; Leonard, A.J.; Shercliff, H.R.; Withers, P.J. Friction stir welding of aluminium alloys. *Int. Mater. Rev.* **2009**, *54*, 49–93. [CrossRef]

26. Çam, G. Friction stir welded structural materials: Beyond al-alloys. *Int. Mater. Rev.* **2011**, *56*, 1–48. [CrossRef]

27. De Filippis, L.; Serio, L.; Palumbo, D.; De Finis, R.; Galietti, U. Optimization and characterization of the friction stir welded sheets of aa 5754-h111: Monitoring of the quality of joints with thermographic techniques. *Materials* **2017**, *10*, 1165. [CrossRef] [PubMed]

28. Roy, G.G.; Nandan, R.; DebRoy, T. Dimensionless correlation to estimate peak temperature during friction stir welding. *Sci. Technol. Weld. Join.* **2006**, *11*, 606–608. [CrossRef]

29. Summers, P.T.; Chen, Y.; Rippe, C.M.; Allen, B.; Mouritz, A.P.; Case, S.W.; Lattimer, B.Y. Overview of aluminum alloy mechanical properties during and after fires. *Fire Sci. Rev.* **2015**, *4*, 1–36. [CrossRef]

30. Liu, K.; Li, X.P.; Rahman, M.; Liu, X.D. Cbn tool wear in ductile cutting of tungsten carbide. *Wear* **2003**, *255*, 1344–1351. [CrossRef]

31. Liu, H.; Hu, Y.; Peng, Y.; Dou, C.; Wang, Z. The effect of interface defect on mechanical properties and its formation mechanism in friction stir lap welded joints of aluminum alloys. *J. Mater. Process. Technol.* **2016**, *238*, 244–254. [CrossRef]

32. Masahashi, N.; Watanabe, S.; Nomura, N.; Semboshi, S.; Hanada, S. Laminates based on an iron aluminide intermetallic alloy and a crmo steel. *Intermetallics* **2005**, *13*, 717–726. [CrossRef]

33. Watanabe, M.; Feng, K.; Nakamura, Y.; Kumai, S. Growth manner of intermetallic compound layer produced at welding interface of friction stir spot welded aluminum/steel lap joint. *Mater. Trans.* **2011**, *52*, 953–959. [CrossRef]

34. Massalski, T.B.; Okamoto, H. *Binary Alloy Phase Diagrams*; ASM International Publishers: Materials Park, OH, USA, 1990.

35. Girard, M.; Huneau, B.; Genevois, C.; Sauvage, X.; Racineux, G. Friction stir diffusion bonding of dissimilar metals. *Sci. Technol. Weld. Join.* **2010**, *15*, 661–665. [CrossRef]

36. Movahedi, M.; Kokabi, A.H.; Reihani, S.M.S.; Najafi, H.; Farzadfar, S.A.; Cheng, W.J.; Wang, C.J. Growth kinetics of Al-Fe intermetallic compounds during annealing treatment of friction stir lap welds. *Mater. Charact.* **2014**, *90*, 121–126. [CrossRef]

37. Picot, F.; Gueydan, A.; Hug, E. Influence of friction stir welding parameters on titanium-aluminum heterogeneous lap joining configuration. *AIP Conf. Proc.* **2017**, *1896*, 050008.

38. Gueydan, A.; Domengès, B.; Hug, E. Study of the intermetallic growth in copper-clad aluminum wires after thermal aging. *Intermetallics* **2014**, *50*, 34–42. [CrossRef]

39. Kidson, G. Some aspects of the growth of diffusion layers in binary systems. *J. Nucl. Mater.* **1961**, *3*, 21–29. [CrossRef]

metals

MDPI

Article

Abnormal Grain Growth in the Heat Affected Zone of Friction Stir Welded Joint of 32Mn-7Cr-1Mo-0.3N Steel during Post-Weld Heat Treatment

Yijun Li [1,2], Ruidong Fu [1,*], Yan Li [1], Yan Peng [2] and Huijie Liu [3]

[1] State Key Laboratory of Metastable Materials Science and Technology, Yanshan University, Qinhuangdao 066004, China; liyijun1987@ysu.edu.cn (Y.L.); 18233566252@163.com (Y.L.)

[2] College of Mechanical Engineering, Yanshan University, Qinhuangdao 066004, China; pengyan@ysu.edu.cn

[3] State Key Lab of Advanced Welding and Joining, Harbin Institute of Technology, Harbin 150000, China; liuhj@hit.edu.cn

* Correspondence: rdfu@ysu.edu.cn; Tel.: +86-335-807-4792; Fax: +86-335-807-4545

Received: 12 March 2018; Accepted: 5 April 2018; Published: 9 April 2018

Abstract: The abnormal grain growth in the heat affected zone of the friction stir welded joint of 32Mn-7Cr-1Mo-0.3N steel after post-weld heat treatment was confirmed by physical simulation experiments. The microstructural stability of the heat affected zone can be weakened by the welding thermal cycle. It was speculated to be due to the variation of the non-equilibrium segregation state of solute atoms at the grain boundaries. In addition, the pressure stress in the welding process can promote abnormal grain growth in the post-weld heat treatment.

Keywords: friction stir welding; abnormal grain growth; high nitrogen steel; post-weld heat treatment; non-equilibrium segregation

1. Introduction

Friction stir welding (FSW) is a solid-state welding technique invented by the Welding Institute in 1991; it was originally used to weld low-melting-point metals such as Al and Al alloys [1]. In recent years, with the development of welding tools, FSW has been applied in high-melting-point metallic materials such as Ti, Zr, and stainless steels [2–4]. FSW has a significant advantage in the welding of high nitrogen steel because metallurgical defects such as nitrogen loss, blowhole defects, hot cracking in the fusion zone, and nitride precipitation in the heat affected zone (HAZ) can be avoided [5]. The grain refinement in the nugget zone (NZ) can improve the strength of the FSW joint, but also lead to a serious decline in plasticity. Post-weld heat treatment (PWHT) can reduce the microstructural gradient of the FSW joint and further improve the plasticity of the joint. Nevertheless, abnormal grain growth (AGG) is found in the HAZ of post-weld heat treated FSW joint, which decreases its yield strength [6].

In the past decades, some scholars have reported that AGG occurred in the NZ of the FSW joint for precipitation-hardened Al alloys after post-weld solution treatment [7–14]. They summarized that factors such as welding parameters [7], texture [8,9], dissolution and growth of precipitations [10,11], localized strain differences [12], non-uniform grain size distribution [13], and the existence of grain boundaries with different mobility [14] may play an important role in this phenomenon.

In comparison, there are few studies investigating the causes of AGG in FSW joint for steels. Only Sun [15] reported that FSW joints for low carbon steel showed AGG in both the NZ and thermal-mechanical affected zone (TMAZ) after annealing for a critical time. They explained that the AGG is caused by the existence of an inhomogeneous strain distribution in the NZ and TMAZ. However, for the HAZ, no large strain existed when compared with the NZ and TMAZ based on the principle of FSW. Thus, the causes for the AGG in the HAZ of FSW joints should be worth a discussion. The current study attempts to find the possible reasons for the AGG in the HAZ of FSW joint for high nitrogen steel after PWHT by physical simulation.

2. Experimental Procedures

Fe-32Mn-7Cr-1Mo-0.3N austenite steel was used in this work. Before the experiment, the test steel was treated with solid solution at 1100 °C for 90 min and then quenched in water to ensure the single austenitic phase in base metal (BM). The plates for FSW were cut to thickness gauge of 3 mm by using wire-electrode cutting. The thermal physical simulated specimens were cut to cylinders with a length and diameter of 12 and 8 mm, respectively, from the BM by using wire-electrode cutting.

A tungsten-rhenium alloy FSW tool, which consisted of a concave shoulder with 16 mm diameter and an unthreaded pin with 3 mm length, was used during the welding process, and the tilting angle of the tool was 2°. A constant rotation speed of 600 rpm and welding speed of 80 mm/min were used under a protective atmosphere of flowing Ar gas. To determine the effect of FSW on the microstructure in the HAZ, the thermal data from this zone were collected using the OMB-DAQ-2416 data acquisition system. The thermocouples were welded in the plates at 10 mm away from the seam center to capture thermal histories. After FSW, PWHT was performed on the as-welded joints by holding at 1100 °C for 90 min and then quenching in water.

The thermal cycle simulation tests were performed on a Gleeble 3500 thermo-mechanical simulator (Dynamic Systems Inc., Austin, TX, USA) at peak temperatures of 450 °C, 550 °C, and 650 °C. Before the tests, the simulated specimens were stuck in the compression anvils. The pressing force is 60 kgf. During the tests, the heating and cooling rates of all specimens were consistent with the actual thermal cycle of the HAZ. It is worth mentioning that there were two load modes of the compression anvils: (1) no additional load mode: during the whole welding thermal cycle simulation, "force" was selected as the mode of compression anvils, and the pressing force was kept at 60 kgf; and (2) with additional load mode: during the heating stage of the welding thermal cycle simulation, "stroke" was selected as the mode of compression anvils, and the displacement was kept at 0 mm. That is to say, the positions of compression anvils were unchanged. Then, during the cooling stage of the simulation process, "force" was selected as the mode of compression anvils, and the pressing force was changed to 60 kgf. After the thermal cycle simulation tests, the heat treatment process, which was the same as the PWHT, was performed on the simulated specimens.

The microstructures of the FSW joints and simulated specimens were observed by optical microscopy (OM) (Carl Zeiss Jena, Oberkochen, Germany) and orientation imaging microscopy (OIM) (HITACHI, Tokyo, Japan). The Vickers hardness profiles of the joints before and after PWHT were measured in the cross-section perpendicular to the welding direction by using of FM-ARS9000 (FUTURE-TECH, Tokyo, Japan).

3. Experimental Results

The transverse cross-section microstructures of the as-welded FSW joint of Fe-32Mn-7Cr-1Mo-0.3N steel without internal defects are presented in Figure 1a. Three typical zones, namely, BM, HAZ, and NZ, are labeled. The BM microstructure in the as-welded joint is shown in Figure 1b, which is mainly composed of grains with an average size of 52 μm and several annealing twins. Figure 1c shows that the NZ grains are significantly refined to an average size of approximately 16 μm, which can be attributed to the dynamic recrystallization [3,16]. No obvious TMAZ with dynamic recovery is found in the FSW joint. The HAZ microstructure in the as-welded joint is shown in Figure 1d,e, which is similar to the features of the BM. In addition, there are some slip bands in the grains of the HAZ, which are typical deformation features. As is well known, in FSW, the HAZ is the zone in which no tool-promoted plastic deformation occurs. This zone should be mainly affected by the welding thermal cycle. However, unlike conventional fusion welding, rigid fixation for welded plates is necessary in the FSW process. Therefore, the plates are not the free ends in the direction perpendicular to the welding.

In the condition of rigid binding of the plates, the material in the HAZ is subjected to the pressure stress caused by the thermal expansion during the FSW process. Given the relatively low temperature, the dynamical recrystallization and dynamical recovery cannot occur in the HAZ. As a result, the cold deformation characteristics such as the slip bands were preserved.

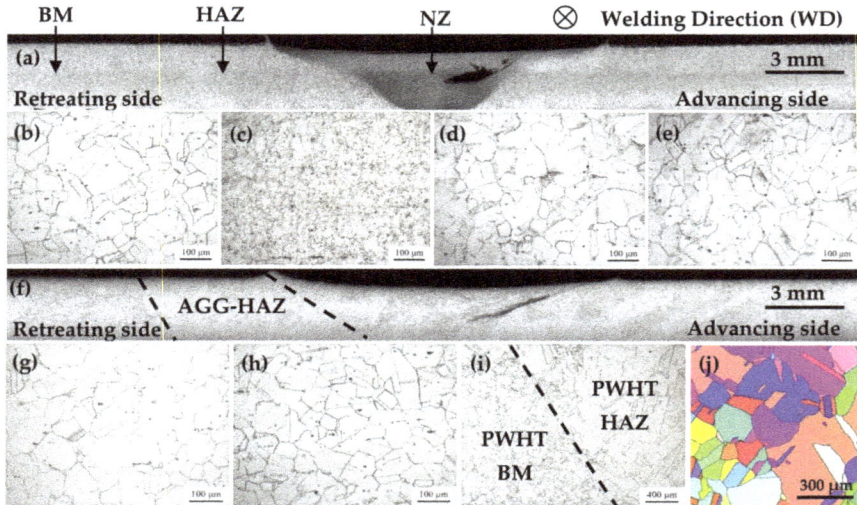

Figure 1. Microstructures in the as-welded and PWHT joints: (**a**) overall cross-section observation of the as-welded joint; (**b**) OM of BM in the as-welded joint; (**c**) OM of NZ in the as-welded joint; (**d**) OM of HAZ in the retreating side of as-welded joint; (**e**) OM of HAZ in the advancing side of as-welded joint; (**f**) overall cross-section observation of the PWHT joint; (**g**) OM of BM in the PWHT joint; (**h**) OM of NZ in the PWHT joint; (**i**) OM of HAZ and BM in the PWHT joint; and (**j**) OIM map of HAZ in the PWHT joint.

The joint overview after PWHT is shown in Figure 1f, and the detailed observations of the regions selected from the joint are shown in Figure 1g–j. The microstructure of the PWHT BM is stable with initial grain size. In comparison, the NZ grains have grown significantly with an average size of approximately 48 μm, which is roughly close to that of the initial BM. Notably, the AGG phenomenon is observed in the HAZ of the PWHT joint. The grains in the AGG regions have clearly grown to an average size of approximately 320 μm as shown in Figure 1i,j. Moreover, the width of these regions are estimated to be approximately 12 mm in each side of the joint, which is wider than the regions characterized by slip bands in the as-welded joint (as labeled by the dotted lines in Figure 1f).

The hardness distributions along the weld cross-section centerline of the as-welded and PWHT joints are shown in Figure 2. In the as-welded joint, the hardness of the NZ is approximately 250 HV, which is much higher than that of the BM (190 HV). In addition, the hardness of HAZ also increases obviously, which should be related to the plastic deformation that occurred in the HAZ. After PWHT, the hardness of the NZ decreases and almost reaches the BM level, which is due to the grain growth in this zone. Besides that, due to the AGG during PWHT in the HAZ, two softened zones appear in the FSW joint. In addition, the width of the softened zones in the PWHT joint is wider than the hardened regions in the as-welded joint.

According to the above analysis, the AGG in the HAZ may be related to the welding thermal cycle and the compressive stress caused by the thermal expansion. To reveal the causes of the AGG in the HAZ, thermal physical simulations with various axial stresses were employed. The black curve in Figure 3 shows the welding thermal cycle experienced by the temperature measuring point set on the

HAZ. The other three curves represent simulated thermal cycles at three different peak temperatures according to the same heating and cooling rate of the actual thermal cycle.

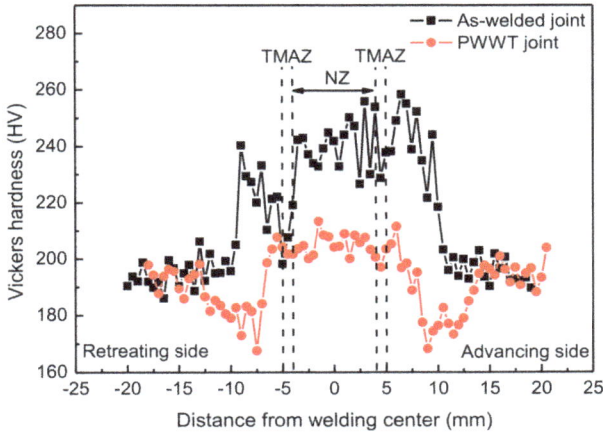

Figure 2. Hardness profile along the transverse direction in the as-welded and PWHT joints.

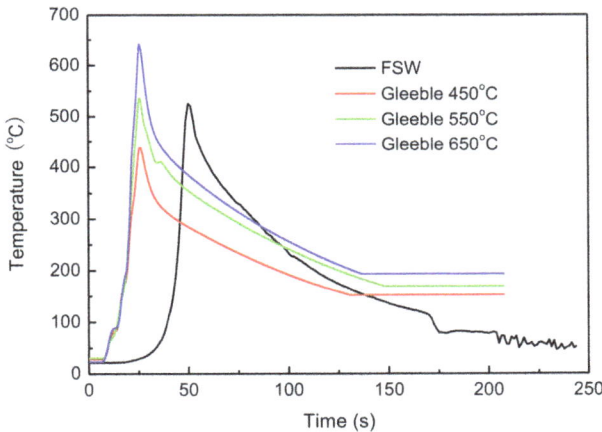

Figure 3. Temperature variation of the HAZ in the FSW joint and the welding thermal simulation specimens.

Figure 4 shows the variation of the axial stresses of the thermal cycle simulated specimen with time under two load modes. The stress of the thermal simulated specimen without additional load is almost unchanged, which is always approximately 10 MPa. In this condition, the specimen is only affected by the thermal cycle. However, during the heating stage of the thermal simulated specimen with additional load, "stroke" was selected as the control mode of compression anvils with a constant displacement of 0 mm. The specimen expands due to the increase in temperature. However, the fixation of the compression anvils' position actually limits the expansion of the thermal simulated specimen. It is equivalent to a compressive load for the specimen. Therefore, as shown in Figure 4, with the increase in temperature, the stress of the specimen with additional load increases significantly. It reaches approximately 175 MPa. In this condition, the specimen is affected by the effects of the thermal cycle and additional load.

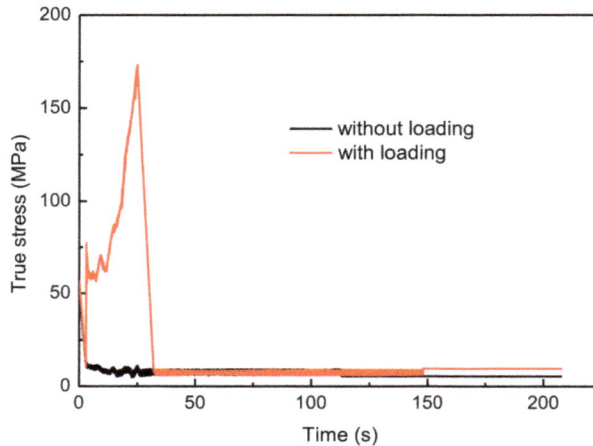

Figure 4. The stress state of the welding thermal simulation specimens.

Figure 5 shows the microstructures of the thermal simulated specimens in the two load conditions when the peak temperature is 550 °C. A comparison of Figures 1 and 5 shows that the microstructure of the simulated specimen without additional load is similar to that of the BM in the as-welded FSW joint. The average grain size is approximately 55 μm, and there is no slip band in the grains. However, the microstructure of the simulated specimen with additional load is similar to that of the HAZ in actual as-weld FSW joint. Moreover, a small number of slip bands can be observed in the grains. This finding indicates that, during the thermal cycle simulated test process, plastic deformation occurred in this specimen due to the effect of the compression anvils.

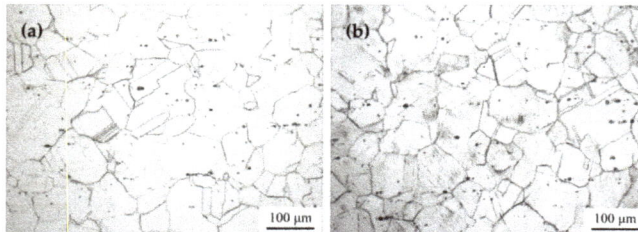

Figure 5. OM micrograph of welding thermal simulation specimens: (**a**) without additional loading; (**b**) with additional loading.

Figure 6 shows the microstructure of the thermal simulated specimens without additional load after heat treatment. First of all, by comparing Figures 5 and 6, the grains of the thermal simulated specimens at three different peak temperatures all grow abnormally after heat treatment. This finding indicates that the microstructural stability of the present steel can be weakened only due to the welding thermal cycle. In addition, it causes AGG in the subsequent heat treatment process. The additional load is not a necessary factor for AGG. Second, the grain size of the three thermal simulated specimens after heat treatment is basically the same, which indicates that the peak temperature of the thermal cycle has a limited influence on the degree of AGG in a temperature range.

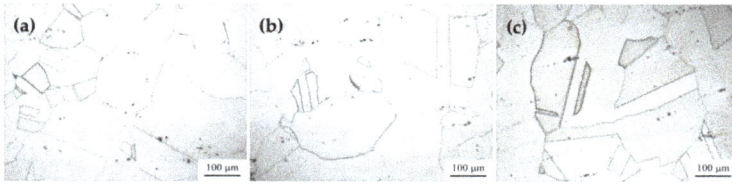

Figure 6. OM micrograph of welding thermal simulation specimens without additional loading after heat treatment: (**a**) 450 °C; (**b**) 550 °C; (**c**) 650 °C.

Figure 7 shows the microstructure of the thermal simulated specimens with additional load after heat treatment. First of all, it is similar to the thermal simulated specimens without additional load. After heat treatment, AGG also occurred in all specimens at three different peak temperatures. Second, compared with the thermal simulated specimens without additional load (by comparing Figures 6 and 7), the grains in the specimens with slight strain are much coarser after subsequent heat treatment. Therefore, this finding proves that, although the additional load is not a necessary factor to induce AGG, it can promote the degree of grain growth in PWHT.

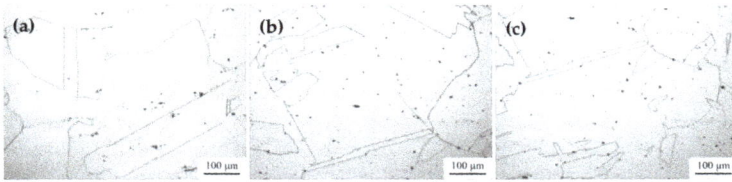

Figure 7. OM micrograph of welding thermal simulation specimens with additional loading after heat treatment: (**a**) 450°C; (**b**) 550°C; (**c**) 650 °C.

4. Discussion

Different from the aged-strengthening Al alloy, there are no second phase particles and other precipitates in the present steel. Moreover, due to the heat treatment before welding, the steel consists of single austenite. Therefore, the AGG in the HAZ of the FSW joint during PWHT can hardly be explained by the dissolution of the second phase particles. In addition, unlike the FSW joint of Al alloy, the AGG was not found in the NZ and TMAZ but the HAZ of the PWHT joint for the present steel. Therefore, the AGG that occurred in the HAZ of the PWHT joint was also not related to the non-uniform grain size distribution. From the previous analysis, the local strain that occurred in the HAZ will promote AGG after heat treatment, but the effect of local strain on the AGG of the HAZ should be discussed. The microstructural stability in terms of the mobility of grain boundaries may account for it.

A study [17] pointed out that, besides the pinning forces due to the second phase particles, the solute atoms in the material will also exert a drag effect on the migration of the grain boundaries. The migration of the grain boundaries will become more difficult if the solute atoms are segregated at the grain boundaries. On the contrary, once the segregation of these atoms disappears or is reduced, the grain boundaries will migrate more easily in the same thermal activation condition, which leads to grain growth.

Generally, the segregation of the solute atoms can be divided into equilibrium segregation and non-equilibrium segregation (NES) [18]. The equilibrium segregation of the solute atoms is a thermal activation process controlled by the solid temperature. Based on the peak temperature of weld thermal cycle shown in Figure 3, it can be confirmed that the equilibrium segregation is not the primary mechanism.

The NES of the solute atoms is a kinetic process and has been wildly accepted in terms of the diffusion of atom-vacant couple induced by the following factors: (1) quenching, (2) radiation, (3) low stress, and (4) recrystallization [17–20]. The NES of solute atoms at grain boundaries arising from quenching effects was first reported by Aust [21] and Anthony [22,23]. The formation of the solute atom-vacancy complex in the matrix is considered to play a major role in the process. When a sample is maintained at a solution treatment temperature for a period of time and then cooled to a certain low temperature, the equilibrium vacancy concentration will be reduced. Consequently, a loss of vacancies will appear along the grain boundaries as they act as a sink of vacancy. The decrease of vacancy concentration causes the dissolution of the complexes into vacancies and solute atoms, which reduces the concentration of complexes near the grain boundaries. Meanwhile, in the regions far away from the grain boundaries, where no other vacancy traps are present, vacancies will recombine with solute atoms to form new complexes, which also reduces the vacancy concentration and increases the complex concentration. This leads to the increase in complex concentration in the regions remote from the grain boundaries. Therefore, a concentration gradient appears between the grain boundaries and the regions beyond it. This gradient drives the complexes to diffuse from the regions far from the grain boundaries to the grain boundaries. It causes the excessive solute atoms to concentrate in the grain boundaries and results in the NES. After segregating at grain boundaries, the solute atoms tend to diffuse into grains due to the concentration gradient between the grain boundary and the inner grain. Thus, the NES of the solute atoms will disappear in the conditions of sufficient annealing and long-term service.

The NES of solute atoms at grain boundaries induced by recrystallization effects was reported by Jahaz [17]. The Cottrell atmosphere and recrystallization phenomena are considered to play a major role in this process. According to the theory of Cottrell atmosphere, the solute atoms tend to gather to the dislocations. Once recrystallization occurs, new grain boundaries move towards high-dislocation-density regions and leave low dislocation densities behind them. Moreover, the existing models for NES consider that the grain boundaries are infinite sinks for vacancies and dislocations. Therefore, with the recrystallization process, the solute atoms have been left at the grain boundaries, resulting in the NES.

In addition, the stress also affects the NES of the solute atoms. Xu [18] pointed out that the grain boundaries emit vacancies when a compression stress is exerted on them and absorb vacancies when a tension stress is exerted. The tensile stress will further promote the NES of the solute atoms at the grain boundaries. On the contrary, the pressure stress will lead to the non-equilibrium dilution of the solute atoms at the grain boundaries.

For high manganese steel and high nitrogen steel, the NES of different solute atoms during quenching process had been previously reported [24,25]. Combined with the above discussions, it can be deduced that the transformation of the NES state of the solute atoms at the grain boundaries leads to the AGG in the HAZ of the joint for the present steel during the whole FSW process and the subsequent heat treatment.

In the heat treatment before welding, the NES of solute atoms at the grain boundaries will occur in the whole plate during the process of water quenching. During the subsequent FSW process, the BM area is not affected by the mechanical and thermal effects of the welding. Moreover, it keeps the NES state of solute atoms at grain boundaries. However, due to the effect of the welding thermal cycle, the NES state of the solute atoms disappears in the HAZ. In the NZ, although the welding thermal cycle will lead to the disappearance of the NES state of solute atoms at the grain boundaries, the recrystallization induces the NES of the solute atoms at the grain boundaries in the NZ again due to the dynamic recrystallization that occurred in the NZ. Therefore, after the FSW, the NES state of solute atoms at grain boundaries exists in the BM and NZ of the as-welded joint. However, the NES state in the HAZ has disappeared. In the HAZ, the drag effect of the solute atoms disappears and thus leads to a significant enhancement in the mobility of the grain boundaries.

During the FSW process, due to the grain refinement in the NZ, the interface energy stored in this zone is relatively high. Therefore, the grains here have the driving force for growth. However, the drag effect of the solute atoms leads to the poor mobility of the grain boundaries. Hence, only normal grain growth has occurred in the NZ during the PWHT. The microstructure of the BM has been in a relatively stable state because of the heat treatment before welding. In addition, the mobility of grain boundaries is poor due to the drag effect of the solute atoms here. Thus, there is no obvious grain growth in the BM of the FSW joint during the PWHT either. After the FSW process, the mobility of grain boundaries is enhanced due to the disappearance of the NES of the solute atoms in the HAZ. As a result, the migration of those grain boundaries becomes easier than in other zones in the joint and thus arouses the AGG during the PWHT. As mentioned in the previous section, due to the rigid restraint produced by the fixture, the HAZ is subjected to pressure stress caused by the thermal expansion during the FSW process. This stress state further aggravates the de-segregation of solute atoms in the HAZ. It further promotes the AGG during PWHT (as simulated in Figures 6 and 7).

5. Conclusions

(1) The AGG occurs in the HAZ of FSW joint for high nitrogen steel after heat treatment at 1100 °C for 90 min and quenching in water. Moreover, the NZ shows a normal grain growth only.

(2) The microstructural stability of high nitrogen steel can be weakened by the welding thermal cycle. It also causes the AGG in the subsequent heat treatment process. In addition, the pressure stress in the welding thermal cycle can promote the AGG in PWHT.

(3) The AGG in the HAZ of FSW joint for high nitrogen steel after PWHT is related to the NES state of solute atoms in the grain boundary. The NES state of solute atoms is altered by the welding thermal cycle. The drag effect of the solute atoms disappears and thus leads to a significant enhancement in the mobility of the grain boundaries. Then, the AGG occurs in the HAZ during the subsequent heat treatment.

Acknowledgments: Financial support by State Key Lab of Advanced Welding and Joining, Harbin Institute of Technology is greatly acknowledged.

Author Contributions: Yijun Li performed research, analyzed the data and wrote the paper; Yan Li helped in the experimental part; Ruidong Fu, Huijie Liu and Yan Peng assisted in the data analysis and revised manuscript.

Conflicts of Interest: The authors declare no conflict of interest.

Abbreviations

FSW	Friction stir welding
HAZ	Heat affected zone
AGG	Abnormal grain growth
NZ	Nugget zone
TMAZ	Thermal-mechanical affected zone
PWHT	Post-weld heat treatment
OM	Optical microscopy
OIM	Orientation imaging microscopy
NES	Non-equilibrium segregation

References

1. Thomas, W.M.; Nicholas, E.D.; Needham, J.C.; Murch, M.G.; Temple-Smith, P.; Dawes, C.J. Friction Stir Welding. Great Britain Patent Application No. 9125978.8, 6 December 1991.

2. Wu, L.H.; Wang, D.; Xiao, B.L.; Ma, Z.Y. Microstructural evolution of the thermomechanically affected zone in a Ti-6Al-4V friction stir welded joint. *Scr. Mater.* **2014**, *78*, 17–20. [CrossRef]

3. Li, Y.J.; Fu, R.D.; Du, D.X.; Jing, L.J.; Sang, D.L.; Zhang, X.Y. Microstructures and mechanical properties of friction stir welded joints of Zr-Ti alloy. *Sci. Technol. Weld. Join.* **2014**, *19*, 588–594. [CrossRef]

4. Chen, Y.C.; Fujii, H.; Tsumura, T.; Kitagawa, Y.; Nakata, K.; Ikeuchi, K.; Matsubayashi, K.; Michishita, Y.; Fujiya, Y.; Katoh, J. Banded structure and its distribution in friction stir processing of 316L austenitic stainless steel. *J. Nucl. Mater.* **2012**, *420*, 497–500. [CrossRef]

5. Du, D.X.; Fu, R.D.; Li, Y.; Jing, L.; Ren, Y.; Yang, K. Gradient characteristics and strength matching in friction stir welded joints of Fe-18Cr-16Mn-2Mo-0.85 N austenitic stainless steel. *Mater. Sci. Eng. A* **2014**, *616*, 246–251. [CrossRef]

6. Li, Y.J.; Fu, R.D.; Du, D.X.; Jing, L.J.; Sang, D.L.; Wang, Y.P. Effect of post-weld heat treatment on microstructures and properties of friction stir welded joint of 32Mn-7Cr-1Mo-0.3N steel. *Sci. Technol. Weld. Join.* **2015**, *20*, 229–235. [CrossRef]

7. Attallah, M.M.; Salem, H.G. Friction stir welding parameters: A tool for controlling abnormal grain growth during subsequent heat treatment. *Mater. Sci. Eng. A* **2005**, *391*, 51–59. [CrossRef]

8. Mironov, S.; Masaki, K.; Sato, Y.S.; Kokawa, H. Relationship between material flow and abnormal grain growth in friction-stir welds. *Scr. Mater.* **2012**, *67*, 983–986. [CrossRef]

9. Mironov, S.; Masaki, K.; Sato, Y.S.; Kokawa, H. Texture Produced by Abnormal Grain Growth in Friction Stir-Welded Aluminum Alloy 1050. *Metall. Mater. Trans. A* **2013**, *44*, 1153–1157. [CrossRef]

10. Aydın, H.; Bayram, A.; İsmail, D. The effect of post-weld heat treatment on the mechanical properties of 2024-T4 friction stir-welded joints. *Mater. Des.* **2010**, *31*, 2568–2577. [CrossRef]

11. Liu, H.J.; Feng, X.L. Effect of post-processing heat treatment on microstructure and microhardness of water-submerged friction stir processed 2219-T6 aluminum alloy. *Mater. Des.* **2013**, *47*, 101–105. [CrossRef]

12. Frigaard, Ø.; Grong, Ø.; Midling, O.T. A process model for friction stir welding of age hardening aluminum alloys. *Metall. Mater. Trans. A* **2001**, *32*, 1189–1200. [CrossRef]

13. Sharma, C.; Dwivedi, D.K.; Kumar, P. Effect of post weld heat treatments on microstructure and mechanical properties of friction stir welded joints of Al-Zn-Mg alloy AA7039. *Mater. Des.* **2013**, *43*, 134–143. [CrossRef]

14. Sato, Y.S.; Watanabe, H.; Kokawa, H. Grain growth phenomena in friction stir welded 1100 Al during post-weld heat treatment. *Sci. Technol. Weld. Join.* **2013**, *12*, 318–323. [CrossRef]

15. Sun, Y.; Fujii, H. Effect of abnormal grain growth on microstructure and mechanical properties of friction stir welded SPCC steel plates. *Mater. Sci. Eng. A* **2017**, *694*, 81–92. [CrossRef]

16. Zhao, Y.; Sato, Y.S.; Kokawa, H.; Wu, A. Microstructure and properties of friction stir welded high strength Fe-36 wt%Ni alloy. *Mater. Sci. Eng. A* **2011**, *528*, 7768–7773. [CrossRef]

17. Jahazi, M.; Jonas, J.J. The non-equilibrium segregation of boron on original and moving austenite grain boundaries. *Mater. Sci. Eng. A* **2002**, *335*, 49–61. [CrossRef]

18. Xu, T.D.; Cheng, B.Y. Kinetics of non-equilibrium grain-boundary segregation. *Prog. Mater. Sci.* **2004**, *49*, 109–208. [CrossRef]

19. Song, S.H.; Zhang, Q.; Weng, L.Q. Deformation-induced non-equilibrium grain boundary segregation in dilute alloys. *Mater. Sci. Eng. A* **2008**, *473*, 226–232. [CrossRef]

20. Song, S.H.; Wu, J.; Wang, D.Y.; Weng, L.Q.; Zheng, L. Stress-induced non-equilibrium grain boundary segregation of phosphorus in a Cr-Mo low alloy steel. *Mater. Sci. Eng. A* **2006**, *430*, 320–325. [CrossRef]

21. Aust, K.T.; Hanneman, R.E.; Niessen, P.; Westbrook, J.H. Solute induced hardening near grain boundaries in zone refined metals. *Acta Metall.* **1968**, *16*, 291–302. [CrossRef]

22. Anthony, T.R. Solute segregation in vacancy gradients generated by sintering and temperature changes. *Acta Metall.* **1969**, *17*, 603–609. [CrossRef]

23. Hanneman, R.E.; Anthony, T.R. Effects of non-equilibrium segregation on near-surface diffusion. *Acta Metall.* **1969**, *17*, 1133–1140. [CrossRef]

24. Tomota, Y.; Strum, M.; Morris, J.W. Microstructural dependence of Fe-high Mn tensile behavior. *Metall. Trans. A* **1986**, *17*, 537–547. [CrossRef]

25. Gavriljuk, V.G.; Berns, H.; Escher, C.; Glavatskaya, N.I.; Sozinov, A.; Petrov, Y.N. Grain Boundary Strengthening in Austenitic Nitrogen Steels. *Mater. Sci. Forum* **1999**, *318–320*, 455–460. [CrossRef]

![metals logo] **metals**

MDPI

Article

Compensation of Vertical Position Error Using a Force–Deflection Model in Friction Stir Spot Welding

Jinyoung Yoon [1,2], Cheolhee Kim [1,3,*] and Sehun Rhee [2]

[1]   Joining Research Group, Korea Institute of Industrial Technology, Incheon 21999, Korea; 0521jin@kitech.re.kr
[2]   School of Mechanical Engineering, Hanyang University, Seoul 04763, Korea; srhee@hanyang.ac.kr
[3]   Department of Mechanical and Materials Engineering, Portland State University, Portland, OR 97201, USA
*   Correspondence: chkim@kitech.re.kr; Tel.: +82-32-850-0222

Received: 27 November 2018; Accepted: 9 December 2018; Published: 11 December 2018

Abstract: Despite increasing need for friction stir spot welding (FSSW) for high-temperature softening materials, system deflection due to relatively high plunging force remains an obstacle. System deflection results in the vertical position error of a welding tool and insufficient plunge depth. In this study, we used adaptive control to maintain plunge depth, the plunging force was coaxially measured, and the position error was estimated using a force–deflection model. A linear relationship was confirmed between the force and deflection; this relationship is dependent on the stiffness of the welding system while independent of process parameters and base materials. The proposed model was evaluated during the FSSW of an Al 6061-T6 alloy sheet and a dissimilar metal combination of Al 6061-T6 alloy/dual phase (DP) 590 steel. Under varying process parameters, the adaptive control maintained a plunge depth with an error of less than 50 μm. Conventional position control has a maximum error of nearly 300 μm.

Keywords: friction stir spot welding; plunge depth; adaptive control; force–deflection model; high-temperature softening materials; dissimilar metal welding

1. Introduction

Friction stir welding (FSW), a form of solid state welding, was developed by The Welding Institute (TWI) of the United Kingdom in 1991 [1]. During FSW, a rotating tool with a pin on its shoulder is inserted into the base material, which is then joined using frictional heat generation and plastic material flow in a solid state. Initially, FSW was mainly applied to aluminum alloys, but its application has been extended to harder metals [2–4]. Successful applications for materials such as copper [5], steel [6,7], titanium alloy [8], and metal matrix composite [9] have been reported. In addition, FSW has been widely accepted as one of the most effective joining processes for dissimilar metal combinations (e.g., Al/Mg [10,11], Al/Fe [12,13], and Al/Ti [14]), for which fusion-welding is challenging.

Friction stir spot welding (FSSW), sometimes called friction stir joining (FSJ), is a variant of traditional FSW [15]. The FSSW process comprises three phases: plunging, bonding, and drawing-out [16]. During the plunging stage, an axial force (the plunging or plunge force) is imposed, and a high-speed turning tool begins to move into the base material until the end of the tool (i.e., the bottom of the pin) reaches a preset plunge depth, where the shoulder of the tool makes contact with and penetrates the upper surface of the base material. Two peaks in the plunging force profile are caused by respective contacts of the pin and shoulder on the base material [17]. The relative motion between base materials and the tool (shoulder and pin surfaces) generates frictional heating; this increases the temperature of the base materials and enhances the plastic flow because Young's modulus and the yield strength of the base materials decrease with increasing temperature. During the bonding stage, the tool position is maintained for a certain duration to attain sufficient heat generation and to stabilize the FSSW process. Finally, during the drawing-out stage, the tool is retracted from the base material [16].

The plunging force is dependent on the temperature and volume of the stir zone, and on the contact area between the tool and the base material. During FSW of high-strength and high-temperature softening materials, the plunging force may exceed the designed limit of the welding system. Excessive plunge force causes a deflection of the welding head and, as a result, insufficient tool plunge depth. Among the three types of commercially available FSW machines (i.e., conventional machine tools, dedicated FSW machines, and industrial articulated robots), dedicated FSW machines have the highest stiffness [18]. However, while minimal, system deflection remains inevitable for high-temperature softening materials.

Previous studies have suggested various methods to compensate for the deflection generated when using nominal position control systems. Smith [19] reported that a constant force control greatly improved weld quality during the lap welding of an Al 6016-T6 alloy with a thickness of 2 mm using a six-axis articulated robot; in that study, force control was essential for robotic FSW in order to compensate for the inherent lack of stiffness. However, plunge depth control was not implemented in the control algorithm, as the author felt that it was more suitable for partial penetration welds or lap welds. In subsequent research [20], temperature was additionally measured using thermocouples embedded into a welding tool, and combined with force control to improve weld quality in partially-penetrated welds. A number of studies have reported using force control techniques, as reviewed by Gibson et al. [21] and Mendes et al. [18], including extensive approaches to enhance the accuracy of the control system [22,23], develop low-cost sensors [24], and establish a kinematic deflection model [25]. Cederqvist et al. [26] suggested depth control for the FSW of copper alloys, for which base materials were diversely manufactured and heat-treated, and the material properties tended to vary widely. Multiple distance sensors, including a laser sensor, a linear variable differential transformer (LVDT), and an axial position sensor, were adopted to achieve consistent plunge depth.

Nevertheless, the methods above have a number of drawbacks. The force control method was devised to overcome inaccuracy in the nominal position control, but constant force control cannot compensate for tool height change caused by the softening of materials according to temperature. For example, where the tool rotation speed increases while holding other parameters fixed, the temperature of the specimen may rise, owing to increased frictional heat generation between the tool and the base material, causing an increase in the plunge depth. During line welding with fixed parameters, the difference in temperature by location can change the plunge depth, even under constant force control. Measuring both the force and the temperature simultaneously can compensate for this; however, a complicated welding tool and head are required. The direct measurement of tool height is an easier way to control plunge depth, but it is hard to apply coaxially, which leads to inaccuracy due to the offset between the tool and sensor positions.

System deflection intrinsically originates from the plunging force through a tool. In this study, adaptive control of the tool height was developed using the relationship between the plunging force and a system deflection during FSSW. The plunging force was coaxially measured using a load cell, and the accuracy of the compensation was investigated for an Al alloy plate and Al/Fe dissimilar metal joint.

2. Materials and Methods

2.1. Experimental Set-Up

The FSSW trials were performed on a 3-axis Cartesian FSW welding system (Hwacheon Machinery, F1300, Gwangju, Korea). This system is a dedicated FSW machine with high stiffness and a zero-degree tilting angle. It has a special interface to correct the vertical position of a welding head using an external signal, which was modified for this study. The tool material was WC-Co12%, and two types of tools were used: a flat shoulder and a conically tapered pin without thread (Figure 1).

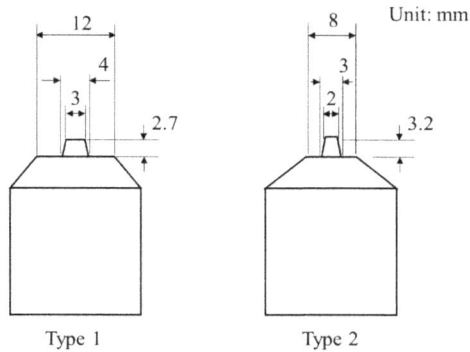

Figure 1. Shapes of welding tools (all dimensions in mm).

The force along the vertical axis was measured using a load cell (Marposs, DDU4, Bentivoglio, Italy) with a resolution of 117 N, and an accuracy of 1.5 kN under a load range of 30 kN. The actual height of the tool was measured for calibration using an LVDT sensor (Marposs, FP50L, Bentivoglio, Italy) with a repeatability of 0.15 μm and a range of 10 mm. The accuracy error of the LVDT sensor varied from 2 μm to 35 μm according to the height of the sensor. The arrangement of the sensing system is shown in Figure 2. The signals from the sensors were collected with a sampling rate of 2 kHz.

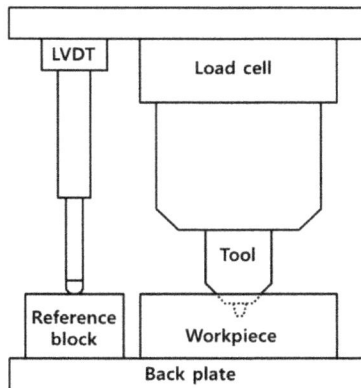

Figure 2. Set-up of the sensing system.

Two kinds of experiments were conducted. First, the force profile during the entire process, as well as the relationship between force and deflection, were identified. The base materials were a 12-mm-thick structural mild steel (SS 400), a 4-mm-thick dual phase high-strength steel (DP 590), and a 25-mm-thick Al alloy sheet (Al 1015). FSSW was performed on one sheet as bead-on-plate (BOP) welding with a type I tool. Prior to welding, the profile of the reference position was measured through dry run welding without a workpiece. After welding, the system deflection was calculated from the position error (i.e., the difference between the reference and measured positions). Using data from various base materials and process parameters, a force–deflection model and adaptive height control algorithm were established.

Second, FSSW using the adaptive height control was implemented to evaluate the developed model. The materials were Al 6061-T6 alloy for BOP welds, with Al 6061-T6 alloy on the top and DP 590 steel on the bottom for dissimilar metal lap welds. The plunging speed was 30 mm/min, and other welding parameters varied, as shown in Table 1.

Table 1. Welding parameters for the experiments.

Experiment No.	Plunge Depth (mm)	Bonding Time (s)	Tool Rotation Speed (rpm)	Tool Shape	Material	Thickness (mm)
1	2.7	6				
2	3	6				
3	3.3	6				
4	3	3	1000			
5	3	6		Type 1	Al 6061-T6	4
6	3	9				
7	3	6				
8	3	6	1500			
9	3	6	2000			
10	3	6	1000			
11	3.2	6		Type 2	Al6061-T6/ DP 590	3 (upper)/2.3 (lower)
12	4	6	1500			

2.2. Force–Deflection Relationship

The profiles of force and actual plunge depth were measured during the entire FSSW process, using a tool with a pin of 2.7 mm in length (Type I), with a tool rotation speed of 500 rpm, a plunging speed of 20 mm/min, and a preset plunge depth of 4 mm. As shown in Figure 3, system deflection initiated as the pin plunged, then increased until the end of the plunging stage. The position error decreased during the bonding stage, owing to greater heat generation and greater plunge depth. While the preset plunge depth was 4 mm, the actual plunge depth was only about 3.1 mm at the end of the bonding stage. During the drawing stage, the position error decreased, then finally disappeared. The pattern of the force profile measured was consistent with that of the position error.

The force–deflection relationship during the plunging stage is shown in Figure 4a. The deflection linearly increased with force in all regions except for the transition zone between ~3 and 4 kN. The relationship was described using Equation 1 and is shown in Figure 4b with the coefficients of determination:

$$Z_d = \begin{cases} 0.0940 + \dfrac{-0.0892}{\left(1+exp^{\left(\frac{x-3.57}{0.158}\right)}\right)+0.0341\cdot x} & \text{for } 0 \leq x < 5 \\ 0.0811 + 0.038\cdot x - 0.000229\cdot x^2 & \text{for } x \geq 5 \end{cases} \tag{1}$$

where Z_d is the deflection in millimeters and x is the force in kilonewtons, the regression was divided into two sections—more than and less than 5 kN—and linear and sigmoid equations were selected as fitting functions.

In order to confirm the force–deflection relationship over an extensive force range, FSSW for different materials (mild steel, high-strength steel, and Al alloy) with various process parameters was conducted. The measured force and position error at the end of the plunging stage are plotted in Figure 5; the relationship was almost perfectly linear with a slope of 0.0323 mm/kN and a coefficient of determination (R^2) of 0.995, regardless of material type or process parameters. This confirms that our force–deflection model was dependent only on the stiffness of the system, but independent of the base materials and process parameters.

Position error due to system deflection was corrected using a feedback control (Figure 6). The feedback system employed a proportional controller to adjust the vertical position using measured force and the force–deflection model. The translation speed of the Z-axis to correct the deflection was programmed according to the amount of position error, as shown in Figure 7. The period of feedback control was 10 µs. Force was measured with a frequency of 2 kHz and an averaged value for each period was used to calculate the deflection. The threshold to initiate feedback control was set to 500 N in order to avoid responding to small and inappropriate disturbances.

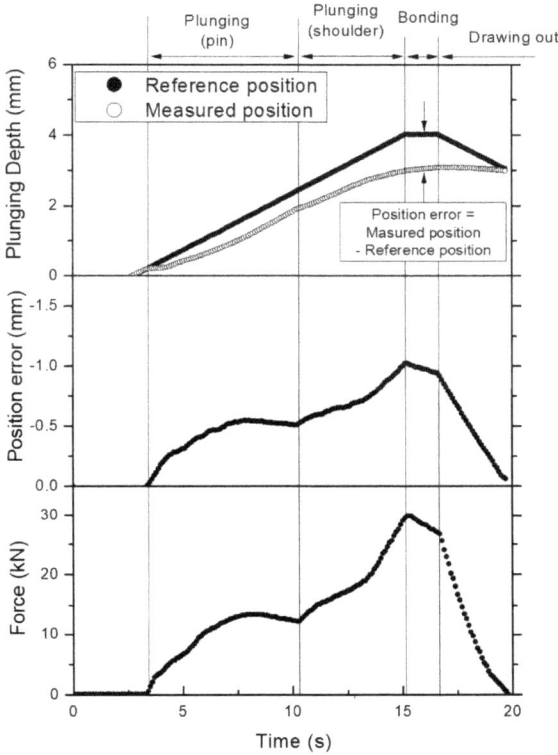

Figure 3. Measured profiles of plunge depth, deflection, and force during friction stir spot welding (FSSW) of steel (SS400).

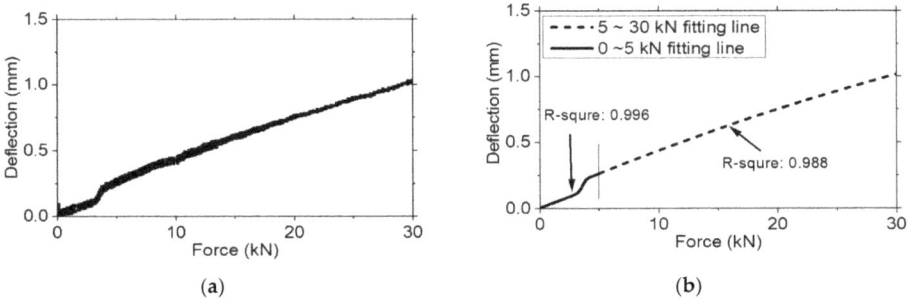

(a)

(b)

Figure 4. Deflection according to force during the plunging stage for the (**a**) measured data and (**b**) fitted results.

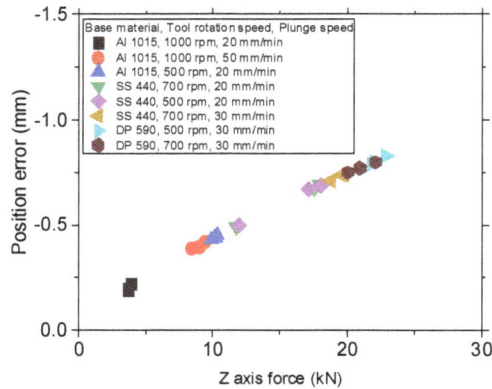

Figure 5. Measured force and deflection at the end of the plunging stage.

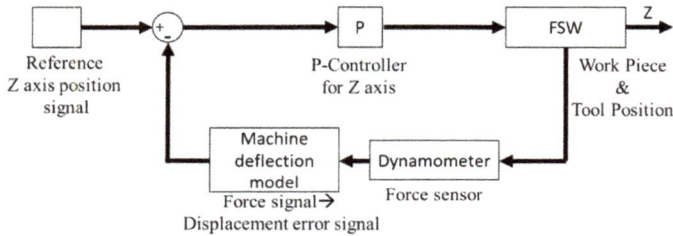

Figure 6. Feedback controller block diagram.

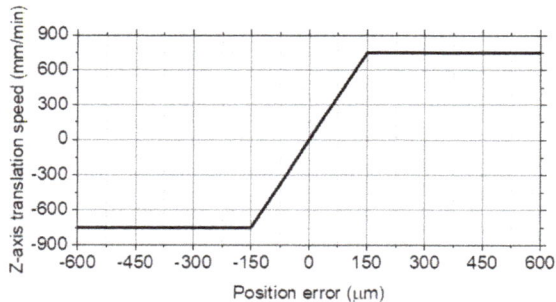

Figure 7. Z-axis translation speed according to position error.

3. Results and Discussion

The feedback system developed was examined under various welding conditions in Table 1 and the fixed plunging speed of 30 mm/min. The plunge depth was measured during the BOP welding of an Al 6061 alloy using parameter set No. 3 (Table 1; a preset plunge depth of 2.7 mm, a bonding time of 6 s, a tool rotation speed of 1000 rpm, and a pin of 2.7 mm in length). During the bonding stage, the feedback control system could follow the reference position with an error of 10 μm (Figure 8). In the early plunging stage, an error of not more than 120 μm was observed, reflecting a measured force that was lower than the threshold force of 500 N, below which the position control did not initiate. Without the control system, the welding tool could not reach the preset plunge depth; the maximum position error was 0.36 mm, sufficient to cause a considerable deficiency in welding quality.

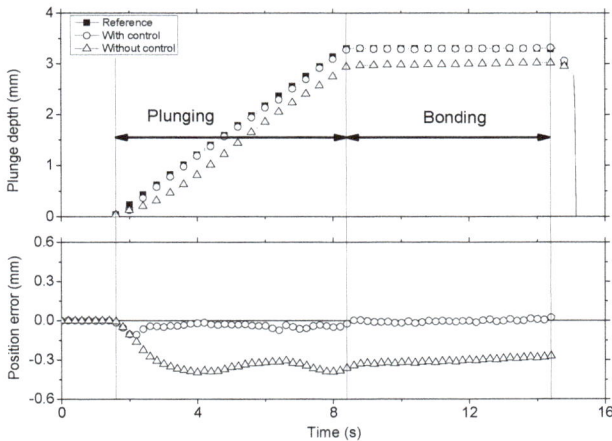

Figure 8. Comparison of plunge depth as a function of process time with and without feedback control (parameter set No. 3; Table 1).

The effect of the process parameters on the plunge depth control was examined to verify the robustness of the control method. First, when the preset plunge depth changed from 2.7 mm to 3.3 mm, the plunge depth with feedback control had an error of 20 μm. This error increased with an increase in the preset plunge depth for the without-control experiment because the increase in plunge depth led to increased force (Figure 9a). Second, when the bonding time changed from 3 s to 9 s, the plunge depth was controlled to within an error of 10 μm (Figure 9b). Without the feedback control, an increase in the bonding time led to less error because the longer bonding time caused higher heat generation and temperature in the welds. However, a longer bonding time (i.e., longer process time) is not preferred for most applications. Third, with respect to the tool rotation speed, the error after the feedback control was less than 50 μm (Figure 9c), slightly higher than that of the preceding two cases, but still acceptable when the accuracy of the sensors and the process characteristics of the welding are considered. As with the longer bonding time, higher rotation speeds caused lower error owing to the higher heat generation and temperature in the welds when no feedback control was used. Finally, the feedback control was applied even when the diameters of the shoulder and pin changed from 12 to 8 mm, and 2.7 to 3.2 mm, respectively. The error after feedback control was less than 40 μm (Figure 9d).

The position control was applied in the FSSW of dissimilar metals (Al 6061-T6 alloy and DP 590 steel) using parameter set No. 12 (Table 1; a preset plunge depth of 4.0 mm, a bonding time of 6 s, a tool rotation speed of 1500 rpm, and a pin of 3.2 mm in length). As shown in Figure 10, during the entire process the maximum position errors for plunge depth were about 100 μm with the control, and 600 μm without the control. The position error at the end of the bond stage reflects the plunge depth error in the final welds, which were 30 μm with the control and 300 μm without the control.

The actual penetration of the tool, as measured in cross-sections of welds, was 3.89 mm with the control and 3.55 mm without the control for a preset plunge depth of 4 mm (Figure 11). The penetration on the cross-sections was slightly lower than the plunge depth measured using the sensor. Tensile-shear tests were implemented for three specimens per case, which were prepared according to ISO 14273:2016. The fracture loads in the tensile-shear test were 3.67 kN for the case with control and 2.41 kN for the case without control. The fracture load increased by 52% by achieving deeper plunge depth using the control.

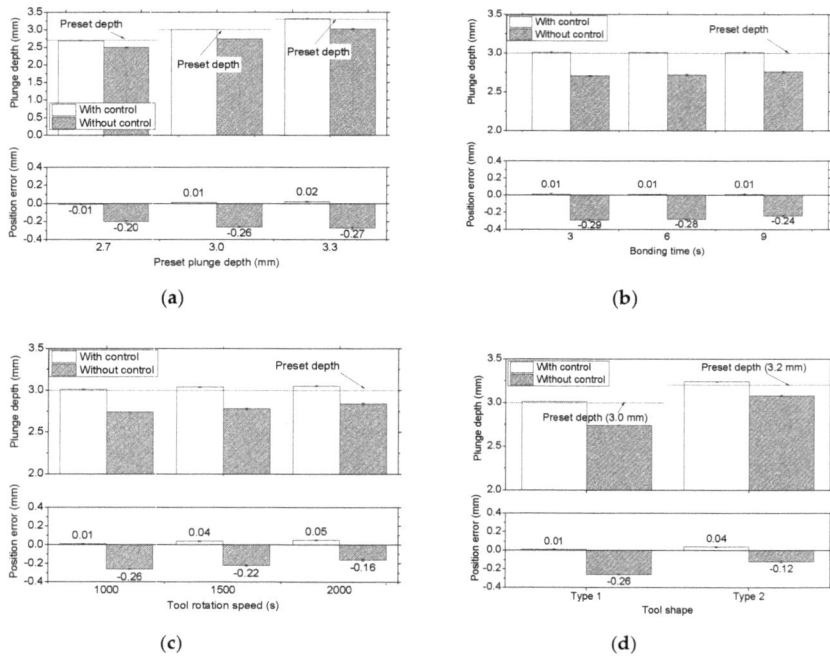

Figure 9. Effect of process parameters (parameter details given in Table 1) for (**a**) preset plunge depth (exp. No. 1–3), (**b**) bonding time (exp. No. 4–6), (**c**) tool rotation speed (exp. No. 7–9), and (**d**) tool shape (exp. No. 10 and 11).

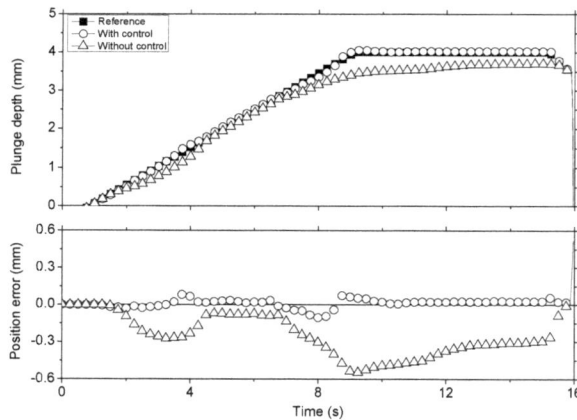

Figure 10. Comparison of the plunge depth as a function of time with and without feedback control (parameter set No. 12; Table 1).

In summary, the vertical deflection of a welding system is linearly proportional to the force for the entire set of base materials and process parameters. Our model, which was established for a specific welding system, could be easily applied without calibration for any combination of material and parameters. During the entire process, the linearity was slightly distorted in the transition region, where the backlash of the system can turn in the opposite direction if the reaction force on the workpiece overcomes the gravity of the welding head. The transition region occurred during the early stages of plunging and did not affect the final position accuracy determined by the plunge depth at the end of the bonding time.

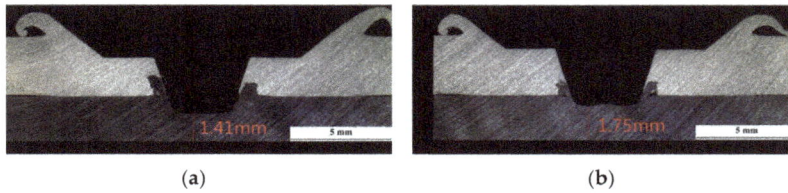

Figure 11. Cross-sections of friction stir spot welding (FSSW) for dissimilar metals (**a**) with and (**b**) without control (plunge depth: 4 mm; stirring time: 6 s; tool rotation speed: 1500 rpm; material: Al 6061-T6 alloy (top) and SPFC 590 dual phase (DP) steel (bottom)).

The constant force control system proposed can improve weld quality; however, it does not consider changes in material properties or process parameters according to temperature, which can modify plunge depth and weld quality. Our force–deflection model-based control can be applied using a coaxial load cell without any auxiliary lateral sensors, and will extend the application of FSSW to high-strength and high-temperature softening materials.

4. Conclusions

This study of FSSW aimed to establish a force–deflection model suitable for use with various materials and parameters, and to implement adaptive control of tool height. Deflections are inherently determined by the force and stiffness of the system. The adaptive control developed here will expand the range of FSSW applications for high-temperature softening materials, and should increase the adoption of articulate robots with flexibility but relatively low stiffness. The main conclusions are as follows:

(1) The deflection of a system is linearly proportional to force and is measurable through a coaxial load cell. The relationship is dependent on nothing but the FSSW system, regardless of the base materials and process parameters (including tool shape).

(2) The performance of the suggested control method was evaluated during the BOP FSSW of an Al alloy. Under varying welding conditions, the position error was corrected to under 50 μm (compared with 0.28 mm when the control was not applied).

(3) The adaptive control for the plunge depth was successfully implemented in the FSSW of an Al/Fe dissimilar metal joint. The welding tool could plunge to the preset depth with an error of 30 μm. In the cross-section of welds, the plunge depth was almost equal to the preset depth, and a sufficiently high hook was formed to ensure the designed joint strength.

Author Contributions: Investigation, J.Y.; Methodology, J.Y. and C.K.; Supervision, C.K. and S.R.; Writing—original draft, J.Y.; Writing—review and editing, C.K.

Funding: This research was supported by the Ministry of Trade, Industry and Energy, Korea.

Acknowledgments: The authors would like to thank Young-Pyo Kim and staff of Hwacheon Machinary for their kind assistance with tailoring the friction stir welding system.

Conflicts of Interest: The authors declare no conflict of interest.

References

1. Thomas, W.; Nicholas, E.; Needham, J.C.; Murch, M.; Templesmith, P.; Dawes, C. Friction Stir Welding. Patent No. PCT/GB92102203, 1 December 1991.
2. Threadgill, P.; Leonard, A.; Shercliff, H.; Withers, P. Friction stir welding of aluminium alloys. *Int. Mater. Rev.* **2009**, *54*, 49–93. [CrossRef]
3. Mishra, R.S.; Ma, Z. Friction stir welding and processing. *Mater. Sci. Eng. R Res.* **2005**, *50*, 1–78. [CrossRef]

4. Nandan, R.; DebRoy, T.; Bhadeshia, H. Recent advances in friction-stir welding–process, weldment structure and properties. *Prog. Mater Sci.* **2008**, *53*, 980–1023. [CrossRef]

5. Lee, W.-B.; Jung, S.-B. The joint properties of copper by friction stir welding. *Mater. Lett.* **2004**, *58*, 1041–1046. [CrossRef]

6. Thomas, W.; Threadgill, P.; Nicholas, E. Feasibility of friction stir welding steel. *Sci. Technol. Weld. Join.* **1999**, *4*, 365–372. [CrossRef]

7. Lienert, T.; Stellwag, W., Jr.; Grimmett, B.; Warke, R. Friction stir welding studies on mild steel. *Weld. J.* **2003**, *82*, 1s–9s. [CrossRef]

8. Ramirez, A.J.; Juhas, M.C. Microstructural evolution in Ti-6Al-4V friction stir welds. *Mater. Sci. Forum* **2003**, *426*, 2999–3004. [CrossRef]

9. Prado, R.; Murr, L.; Shindo, D.; Soto, K. Tool wear in the friction-stir welding of aluminum alloy 6061+Al$_2$O$_3$: A preliminary study. *Scr. Mater.* **2001**, *45*, 75–80. [CrossRef]

10. Hirano, S. Microstructure of dissimilar joint interface of magnesium alloy and aluminum alloy by friction stir welding. *Q. J. Jpn. Weld. Soc.* **2003**, *21*, 539–544. [CrossRef]

11. Lee, W.-B.; Schmuecker, M.; Mercardo, U.A.; Biallas, G.; Jung, S.-B. Interfacial reaction in steel–aluminum joints made by friction stir welding. *Scr. Mater.* **2006**, *55*, 355–358. [CrossRef]

12. Kimapong, K.; Watanabe, T. Friction stir welding of aluminum alloy to steel. *Weld. J.* **2004**, *83*, 277s–282s.

13. McLean, A.; Powell, G.; Brown, I.; Linton, V. Friction stir welding of magnesium alloy AZ31B to aluminium alloy 5083. *Sci. Technol. Weld. Join.* **2003**, *8*, 462–464. [CrossRef]

14. Chen, Y.C.; Nakata, K. Microstructural characterization and mechanical properties in friction stir welding of aluminum and titanium dissimilar alloys. *Mater. Des.* **2009**, *30*, 469–474. [CrossRef]

15. Yang, X.W.; Fu, T.; Li, W.Y. Friction Stir Spot Welding: A Review on Joint Macro- and Microstructure, Property, and Process Modelling. *Adv. Mater. Sci. Eng.* **2014**, *2014*. [CrossRef]

16. Nguyen, N.-T.; Kim, D.-Y.; Kim, H.Y. Assessment of the failure load for an AA6061-T6 friction stir spot welding joint. *Proc. Inst. Mech. Eng. Pt. B J. Eng. Manuf.* **2011**, *225*, 1746–1756. [CrossRef]

17. Gerlich, A.; Su, P.; North, T.H. Tool penetration during friction stir spot welding of Al and Mg alloys. *J. Mater. Sci.* **2005**, *40*, 6473–6481. [CrossRef]

18. Mendes, N.; Neto, P.; Loureiro, A.; Moreira, A.P. Machines and control systems for friction stir welding: A review. *Mater. Des.* **2016**, *90*, 256–265. [CrossRef]

19. Smith, C.B. Robotic friction stir welding using a standard industrial robot. In Proceedings of the Second Friction Stir Welding International Symposium, Gothenburg, Sweden, 26–28 June 2000; TWI Ltd.: Cambridge, UK, 2000.

20. Fehrenbacher, A.; Smith, C.B.; Duffie, N.A.; Ferrier, N.J.; Pfefferkorn, F.E.; Zinn, M.R. Combined Temperature and Force Control for Robotic Friction Stir Welding. *J. Manuf. Sci. Eng.* **2014**, *136*. [CrossRef]

21. Gibson, B.T.; Lammlein, D.H.; Prater, T.J.; Longhurst, W.R.; Cox, C.D.; Ballun, M.C.; Dharmaraj, K.J.; Cook, G.E.; Strauss, A.M. Friction stir welding: Process, automation, and control. *J. Manuf. Process.* **2014**, *16*, 56–73. [CrossRef]

22. Longhurst, W.R.; Strauss, A.M.; Cook, G.E.; Cox, C.D.; Hendricks, C.E.; Gibson, B.T.; Dawant, Y.S. Investigation of force-controlled friction stir welding for manufacturing and automation. *Proc. Inst. Mech. Eng. Pt. B J. Eng. Manuf.* **2009**, *224*, 937–949. [CrossRef]

23. Guillo, M.; Dubourg, L. Dual control loop force/position with secondary encoders: Impact & improvement of industrial robot deviation on FSW quality. In Proceedings of the 11th International Symposium on Friction Stir Welding, Cambridge, UK, 17–19 May 2016.

24. Gibson, B.T.; Cox, C.D.; Longhurst, W.R.; Strauss, A.M.; Cook, G.E. Exploiting robotic link deflection for low-cost force measurement in manufacturing. *Measurement* **2012**, *45*, 140–143. [CrossRef]

25. De Backer, J.; Bolmsjö, G. Deflection model for robotic friction stir welding. *Ind. Robot. Int. J.* **2014**, *41*, 365–372. [CrossRef]

26. Cederqvist, L.; Garpinger, O.; Nielsen, I. Depth and Temperature Control During Friction Stir Welding of 5 cm Thick Copper Canisters. In *Friction Stir Welding and Processing IX*; Hovanski, Y., Mishra, R., Sato, Y., Upadhyay, P., Yan, D., Eds.; Springer: Cham, Switzerland, 2017; pp. 249–260.

metals

MDPI

Article

Another Approach to Characterize Particle Distribution during Surface Composite Fabrication Using Friction Stir Processing

Namrata Gangil [1], Sachin Maheshwari [1], Emad Abouel Nasr [2,3,*], Abdulaziz M. El-Tamimi [2], Mohammed A. El-Meligy [4] and Arshad Noor Siddiquee [5,*]

[1] Division of Manufacturing Processes and Automation Engineering, Netaji Subhas Institute of Technology, New Delhi 110078, India; namrata.gangil@gmail.com (N.G.); ssaacchhiinn@gmail.com (S.M.)
[2] Industrial Engineering Department, College of Engineering, King Saud University, Riyadh 11421, Saudi Arabia; atamimi@ksu.edu.sa
[3] Mechanical Engineering Department, Faculty of Engineering, Helwan University, Cairo 11792, Egypt
[4] Advanced Manufacturing Institute, King Saud University, Riyadh 11421, Saudi Arabia; melmeligy@ksu.edu.sa
[5] Department of Mechanical Engineering, Jamia Millia Islamia, New Delhi 110025, India
* Correspondence: eabdelghany@ksu.edu.sa (E.A.N.); arshadnsiddiqui@gmail.com (A.N.S.); Tel.: +966-569-958-202 (E.A.N.); +91-986-871-8018 (A.N.S.)

Received: 4 July 2018; Accepted: 19 July 2018; Published: 24 July 2018

Abstract: Surface composite fabrication through Friction Stir Processing (FSP) is evolving as a useful clean process to enhance surface properties of substrate. Better particle distribution is key to the success of surface composite fabrication which is achieved through multiple passes. Multiple passes significantly increase net energy input and undermine the essence of this clean process. This study proposes a novel approach and indices to relate the particle distribution with the FSP parameters. It also proposes methodology for predicting responses and relate the response with the input parameter. Unit stirring as derived parameter consisting of tool rotation speed in revolutions per minute (rpm), traverse speed and shoulder diameter was proposed. The particle distribution was identified to be achieved in three stages and all three stages bear close relationship with unit stirring. Three discrete stages of particle distribution were identified: degree of spreading, mixing and dispersion. Surface composite on an aerospace grade aluminum alloy AA7050 was fabricated successfully using TiB_2 as reinforcement particles. FSP was performed with varied shoulder diameter, rotational speed and traversing speed and constant tool tilt and plunge depth using single pass processing technique to understand the stages of distribution. Significant relationships between processing parameters and stages of particle distribution were identified and discussed.

Keywords: friction stir processing; aluminum alloy; surface composites; particle distribution

1. Introduction

Surface composite (SC) fabrication via friction stir processing (FSP) has become popular over the last decade as it is a clean process and capable of developing superior microstructure and properties. Apart from other materials, aluminum alloys are largely used for SC fabrication by employing various ceramics/hard particle(s) in powder form to provide reinforcement in the ductile interior [1,2]. There is a constant urge to improve mechanical properties such as hardness, wear resistance and corrosion resistance in general and strength in particular, in aluminum alloys to further enhance its high specific strength and other associated mechanical properties. Age-hardenable aluminum alloys such as 7xxx series alloys are among the strongest aluminum alloys. Surface composite fabrication may be an attractive approach to further strengthen these alloys. In precipitation-hardenable alloys, aging imparts maximum allowable strength and aging is the final treatment. Any further treatment of age-hardened alloys usually leads to drop in strength due to Ostwald ripening. If suitable treatment including SC fabrication coupled with/without another stage of aging can provide further improvement in the strength, it may lead to great weight savings. This is the main driving force for the present work in which single pass FSP was employed to fabricate SCs on 7xxx series aluminum alloy. The FSP is recognized as a clean process mainly because it is a solid state fabrication technology and it retains the value-additions of primary processing. In addition, it uses significantly less energy, and the energy being supplied is fruitfully utilized in material properties enhancement. The process is also free from effluent emanation.

To fabricate SCs, the reinforcement particles are first preplaced in the matrix material (i.e., the base material) through a special FSP tool. The tool, while rotating, is inserted in the base material (BM), and after insertion it is traversed in the processing direction (as shown in Figure 1). Friction between the tool's shoulder and BM surface generates heat, which softens the material which is present under the shoulder. The rotation coupled with traversing action of the tool mixes preplaced reinforcement particles in the matrix material through stirring action. During stirring, the stirred material is consolidated at the trailing edge of the tool to create processed zone. In this way, the entire surface can be processed by rastering the tool translation [3,4].

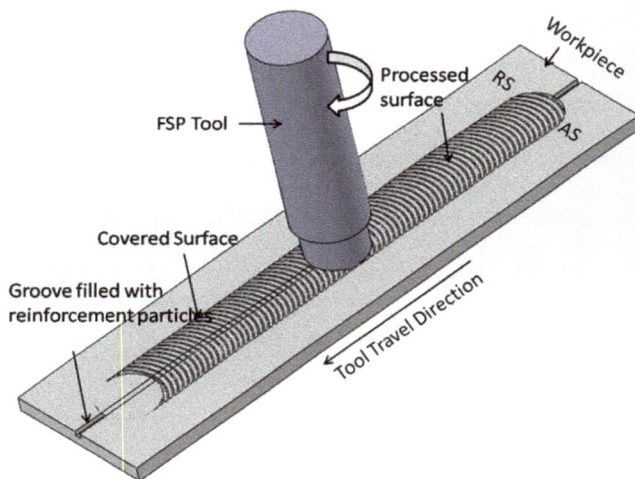

Figure 1. Schematic diagram of friction stir processing.

7xxx (Zn-Mg-Cu Al-alloys) is an age-hardenable high strength aluminum alloys series. It is extensively employed in structural application in aerospace, aircraft and military sectors [2,5]. These alloys are commonly used in T6 and T7XX treated conditions. The T6 treatment imparts peak strength and T7XX treatment imparts high resistance to stress corrosion cracking (SCC) while simultaneously sacrificing some strength [6,7]. AA7050-T7451 aluminum alloy typically finds applications in fuselage frames, bulkheads, wing skins, etc. due to its good strength, toughness and crack resistance [6,8]. The improvement in mechanical properties in SCs also depends very strongly on factors such as grain refinement and homogeneity of distribution of hard phase. In the case of age-hardened alloys, the enhancement of strength is even more difficult. This is mainly because the strengthening imparted by the minute, dense and homogenously distributed hardening precipitates overplays all other strengthening mechanisms. SC fabrication via FSP produces ultrafine grains and distributes reinforcement particles in the processed region. Improvement in properties in age-hardened alloys can be achieved only when distribution of particles is achieved without over-aging; otherwise, it may even result in the drop in strength. Thus, SC fabrication of high strength precipitation hardened 7xxx alloys is challenging. Furthermore, limited literature is available on surface composite fabrication through FSP on 7xxx series of aluminum alloys [2,9–12].

Although much work on SC fabrication in the area of reinforcement particle distribution has been reported, most studies report parametric investigation and effect of multiple passes. Furthermore, most studies performed on SC fabrication on Al-alloys are on non-age-hardenable alloys. Incidentally, multi-pass FSP may not be an effective strategy for strengthening of age-hardened alloys as heat input during every pass promotes chances of ripening. Understanding of the mechanism of distribution of hard phase in the substrate is still evolving. The dispersion and distribution of particles greatly depends on the effectiveness of stirring action. Under conditions of lack of homogeneity of distribution, considerable portion of packed particles remains accumulated at/near original pre-placement. Particle accumulated region(s) act as discontinuities which share little in-service load and reduce the effective load bearing area of the entire cross-section. It also acts as stress raiser and leads to stress concentration. Moreover, such in-homogeneities make the testing of fabricated SCs more complex, as basic principles of testing require material to be homogeneous.

Discontinuities caused by particle accumulation produce stress concentration in the vicinity of discontinuity. Exact theory of mechanics shows that the peak intensity of stress concentration "σ_{max}" exists (as shown in Figure 2) at the extremities of the discontinuity and it can be as high as three times the value of nominal stress "σ_n" for a circular discontinuity. For a noncircular discontinuity (such as the elliptical one typically shown in the figure), magnitude of the stress concentration depends on the length of major and minor axes. The value of "σ_{max}" can be much higher for elliptical discontinuities or for those with sharp corners [13–15]. In any case, a sudden discontinuity adversely affects the strength in elastic range and when the part is subjected to variable loads. If a small discontinuity in the form of accumulation exists, the fabricated SC may be weakest in the vicinity of discontinuity and may fail even though rest of the material may be much stronger. The homogeneous particle distribution without accumulation is, thus, most important.

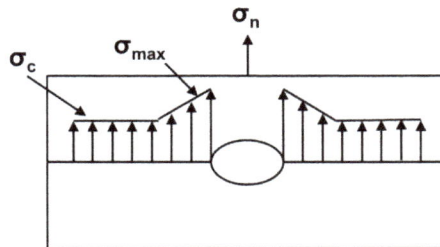

Figure 2. Stress concentration due to elliptical discontinuity.

Interestingly, whereas the material movement has been considerably investigated for friction stir welding (FSW) [16], the same is not true for SC fabrication via FSP [9,17,18]. Material movement, in the case of SC fabrication, is quite different because of heterogeneous makeup of the material being stirred. To obtain better particle distribution, researchers have used multi-pass FSP as a general strategy [9,17–19]. However, multiple passes raise the heat input of processed zone (PZ) which adds to over-aging woes, aiding solutionizing and coagulation of precipitates, which may drastically reduce the strength of age-hardened matrix based SCs. The increased net energy input due to multiple passes also undermines the very essence of energy efficacy of this clean process. This strategy is also time and energy consuming and often may not yield desired enhancements in properties of age-hardened alloys [20]. Thus, use of multi-pass as a general strategy to obtain homogeneous distribution may not be a wise alternative, especially for age-hardened alloys. Instead, efforts should be directed to evolve an understanding on exactly how major process parameters engage with the particle distribution in the material being processed. Studies of such kind are scarcely reported. Present article makes an attempt to investigate the manner in which FSP process parameters engage with the heterogeneous mix of material (age-hardened BM and reinforcement particles) being stirred through single pass FSP. In addition, since the individual FSP parameters may have contradicting effect of the response, a unique compound input parameter represented as "unit stirring", which relates to responses, is defined. A new three stage particle distribution and novel indices for measures of effectiveness of these stages in single pass FSP is also proposed. Results and inferences are demonstrated through fabrication of AA7050/TiB$_2$ SC.

2. Materials and Methods

Aerospace grade AA7050-T7451 (aerospace materials specification: AMS 4342) aluminum samples having dimension of 170 mm × 85 mm × 6 mm were used as BM. Chemical composition of as-received BM is given in Table 1. Grooves of 1.5 mm in depth and 1.5 mm wide were made on the surface of plates. Titanium di-boride (TiB$_2$) powder was used as reinforcement. A pin-less tool was employed to initially cover and compact the grooves filled with TiB$_2$ particles. High carbon high chromium (HCHCr) steel tool (Figure 3) having anti-clock wise scrolled shoulder (0.75 mm width and 0.5 mm height of the scroll), cylindrical pin 6.5 mm in diameter, 3 mm in length was selected based on our previous study [21] and employed for processing. The FSP was performed in the position control mode with total plunge depth of 3.2 mm and a tool tilt of 2° was used.

Table 1. Chemical composition of AA7050-T7451 (wt %).

Element	Cu	Mg	Zn	Fe	Mn	Si	Cr	Ti	Zr	Al
AA7050	2.2	2.3	6.2	0.07	0.01	0.03	0.01	0.06	0.1	Remainder

The FSP was performed on an indigenously developed FSW/FSP setup. The SCs were fabricated through a series of experiments comprising varying combinations of FSP parameters (as given in Table 2) with single-pass processing. The newly defined unit stirring is estimated and presented in Table 2. Experiments in three replicates were performed and average response was considered for analysis.

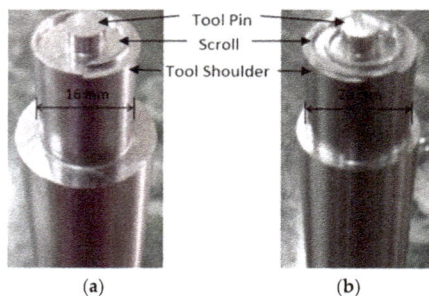

Figure 3. FSP tools having diameter (**a**) 16 mm, and (**b**) 20 mm.

After FSP, microstructural analysis of PZ was carried out for which specimens were prepared using standard metallographic procedure. The metallographic samples were etched with modified Keller's reagent (150 mL distilled water, 6 mL HCL, 6 mL HF, and 3 mL HNO$_3$) for 10 s. Macroscopic images were taken using Stereo-zoom microscope (Focus, Japan). Microstructural observations were carried out by employing computer interfaced optical microscopy (QS Metrology, India).

The areas of characteristic regions were measured using the microstructural image analysis system embedded software.

It is important to note that the unit stirring is representative of the rate of processing too. If the processing rate is high, stirring rate will be less and vice versa. Moreover, under prevailing conditions of FSP the processing rate in some case(s) (represented by unit stirring referenced in Table 2) typically may be as fast as 0.007 mm/rev-shoulder diameter, or in some cases may be as slow as 0.00223 mm/rev-shoulder diameter. It is imperative that the effects of each parameter on every individual response are different and sometimes contradicting too. To interpret the effect of parameters in a meaningful way, a composite parameter, unit stirring "ω", is derived, which comprises three main parameters: rotational speed, traverse speed and shoulder diameter. Unit stirring is expressed as traverse speed over tool rotation and shoulder diameters; and it is representative of axial processing rate per unit rpm and per unit shoulder diameter. For the age-hardened 7050-T7451, the distribution of particle in single pass FSP without defect formation (e.g., tunneling, voids and excessive flash) is very difficult. Further, a very wide range of main FSP parameters, especially shoulder diameter, may not be feasible to employ. Under these circumstances, a single compound parameter may prove to be useful, as it takes care of contradicting effect of individual FSP parameters.

Table 2. Experiment plan, processing parameters and derived parameters.

Sample No.	Processing Parameters			Unit Stirring $\omega=(\frac{T}{R \times SD})$
	Shoulder Diameter (SD) mm	Tool Rotation (R) rpm	Traverse Speed (T) mm/min	
1	16	710	50	0.00440
2	16	710	80	0.00704
3	16	1120	50	0.00279
4	16	1120	80	0.00446
5	20	710	50	0.00352
6	20	710	80	0.00563
7	20	1120	50	0.00223
8	20	1120	80	0.00357

3. Results and Discussion

Simultaneous tool rotation and traversing is responsible for stirring and net material movement. SC processing rate can be assessed in terms advancement of tool per revolution and can be estimated by the ratio of traversing speed "T" to tool rotation "R" (i.e., T/R). Typically, in simple FSP (i.e., without reinforcement), tool moves the material ahead layer by layer, and deposits it behind to replenish the space created by the advancing tool. This layer by layer movement is accomplished through a series of stick and slip actions between the layers and tool. However, in the case of SC fabrication, the reinforcement is in the form of a mix of matrix material and unbinded, loose and discrete particles. Thus, the movement of unbinded particle may be sluggish, whereas the matrix material may move on. How sluggish is the movement of reinforcement particles, however, depends on several factors such as shape and size of particles, friction characteristics, conditions of temperature, flow stress, etc. Given this situation, if the processing rate is high, it may be possible that the reinforcement particles may slip more, stick less and hence may not move with the same pace as that of the matrix material. At slower processing rate, the material will be stirred more as number of unit traversing of tool per unit revolution will be less. Under the conditions of prevailing processing rates, by the time the shoulder traverses a distance equal to shoulder diameter, the reinforcement particles may have been subjected to several repeated actions of stirring, causing the particles to first spread, then mix and finally disperse in fabricated SCs.

If reinforcement particles did not remain accumulated in place but moved effectively, homogeneous distribution can occur in three consecutive stages: (a) spreading; (b) mixing; and (c) even or homogeneous distribution or dispersion. During spreading, the particles get scattered (yet remain accumulated closely) without adequate mixing with BM. In mixing stage, particles, although mix with BM, remain too closely packed and reinforcement–BM bonding is weak. In final stage, the particles become intimately mixed with BM as well as well dispersed, have intimate bonding with the matrix and the mixture is even and homogeneous.

Depending on processing rate, regions that undergo these stages of distribution are formed and are schematically shown in Figure 4. The outline encompassing all zones is the entire processed zone (PZ, with the area of PZ specified as A_{pz}) which has been processed by the tool. There also exists region where no reinforcement could reach. This region, however, has been subjected to simple FSP and undergoes grain refinement. Attempts of tool during stirring subject the particles, in succession, to spreading, portions of spread particle to mixing and mixed particles to dispersion. In accumulation (or spreading) region, the particles do not undergo subsequent stages of mixing/dispersion, because tool's sustained efforts were not available as it has moved ahead without adequate stirring leaving behind particles to remain spread only. Subsequent actions of tool in the mixed regions actually result in increasing the extent of spreading, causing more even distribution.

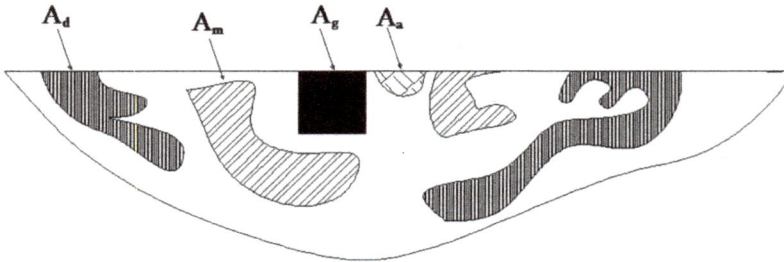

Figure 4. PZ with regions of accumulated (shown as A_a), mixed (A_m) and dispersed (A_d) reinforcement. A_g is the initial grooved in which reinforcement was packed.

Ideally, for perfectly dispersed SC, there should be no accumulation and entire PZ should have homogenous distribution of reinforcement. In practical situation, some regions may always exist where no reinforcement particle is present, although this region undergoes simple FSP. In the present investigation, these regions were visible. The areas of each region were measured and their values are shown in Table 3.

Table 3. Processed zone (PZ) dimensions.

S. No.	Unit Stirring ω	Unreinforced FSPed Area ($A_u = A_{pz} - A_r$) (mm^2)	Area of PZ (A_{pz}) (mm^2)	Area of Accumulation (A_a) (mm^2)	Area of Mixing (A_m) (mm^2)	Area of Dispersion (A_d) (mm^2)	Reinforcement Area, A_r (=A_a + A_m + A_d) (mm^2)
1	0.00440	8.18	16.89	0.47	1.11	7.13	8.71
2	0.00704	8.14	18.25	0.31	0.92	8.88	10.11
3	0.00279	13.03	20.99	0.32	1.97	5.67	7.96
4	0.00446	8.77	15.29	0.44	1.79	4.29	6.52
5	0.00352	8.65	19.96	0.59	0.39	10.33	11.31
6	0.00563	5.31	21.08	0.58	0.2	14.99	15.77
7	0.00223	8.55	20.52	0.06	0.87	11.04	11.97
8	0.00357	9.81	18.33	0.54	0.69	7.29	8.52

The effects of derived parameter "ω" investigated for two different diameters, i.e., 16 mm and 20 mm, and areas of zones pertaining to characteristic stages have been plotted and the same is shown in Figure 5. The results show that the nature of plots of same response for the different diameters possesses good symmetry.

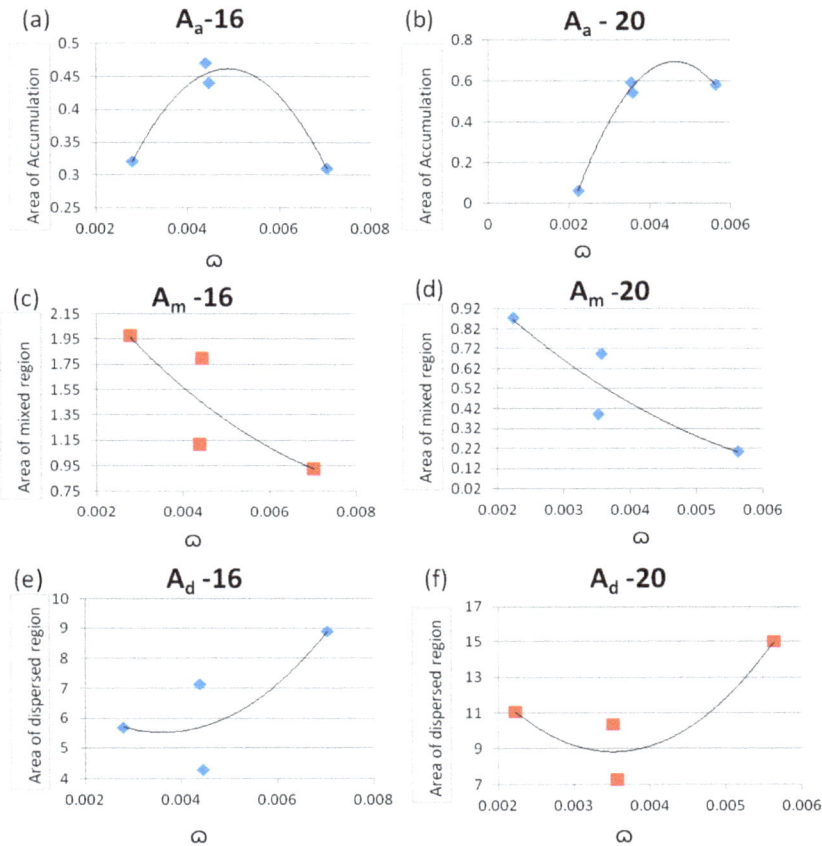

Figure 5. Effect of "ϖ" on (**a**,**b**) A_a, (**c**,**d**) A_m, and (**e**,**f**) A_d for 16 and 20 mm shoulder diameter, respectively.

This is to note that the most desirable response is the area of dispersion zone (i.e., A_d as already defined in Figure 4) and it follows an increasing trend while rest of the two responses follow a decreasing trend with the increase of the value of "ϖ". This is a useful inference to attain high degree of distribution in a single pass FSP operation during SC fabrication.

Further, the estimation of extent of particle distribution during each stage can be accomplished by three indices: degree of accumulation (D_a), degree of mixing (D_m) and degree of dispersion (D_d). D_a is proposed to be defined, with respect to area of cross section of groove (A_g) in which the powder was initially packed, through a function given in Equation (1):

$$D_a = \frac{A_a}{A_g} \times 100\% \tag{1}$$

where A_a is area of accumulated region.

During SC fabrication accumulation should not occur; the mixing regions is undesirable and should be as small as possible; and dispersion should be maximum. Thus, the proposed D_m is defined as given in Equation (2):

$$D_m = \frac{A_m}{A_p + A_a} \times 100\% \tag{2}$$

where A_p is the area of simple FSPed region where no reinforcement has taken place. The proposed D_d is defined as per function given in Equation (3):

$$D_d = \frac{A_d}{A_p + A_a + A_m} \times 100\% \tag{3}$$

Typically, the extent of all three indices bears close relationship with FSP process parameters. The values of these indices are estimated and are given Table 4.

Table 4. Effect of derived parameters on the stages of particle distribution.

S. No.	Unit Stirring	Degree of Accumulation, D_a	Degree of Mixing, D_m	Degree of Dispersion D_d
		Shoulder diameter: 16 mm		
1	0.00440	20.89	12.83	73.05
2	0.00704	13.78	10.89	94.77
3	0.00279	14.22	14.76	37.01
4	0.00446	19.56	19.44	39.00
		Shoulder Diameter: 20 mm		
5	0.00352	26.22	4.22	107.27
6	0.00563	25.78	3.39	246.14
7	0.00223	2.67	10.10	116.46
8	0.00357	24.00	6.67	66.03

The estimated indices are also plotted against unit stirring for the two selected shoulder diameters, i.e., 16 and 20 mm. These plots are given in Figure 6. Results from the experiments were also analyzed through macro- and micrographs taken from each region to corroborate the morphology and particle densities and pattern of distribution in the characteristic stages of distribution.

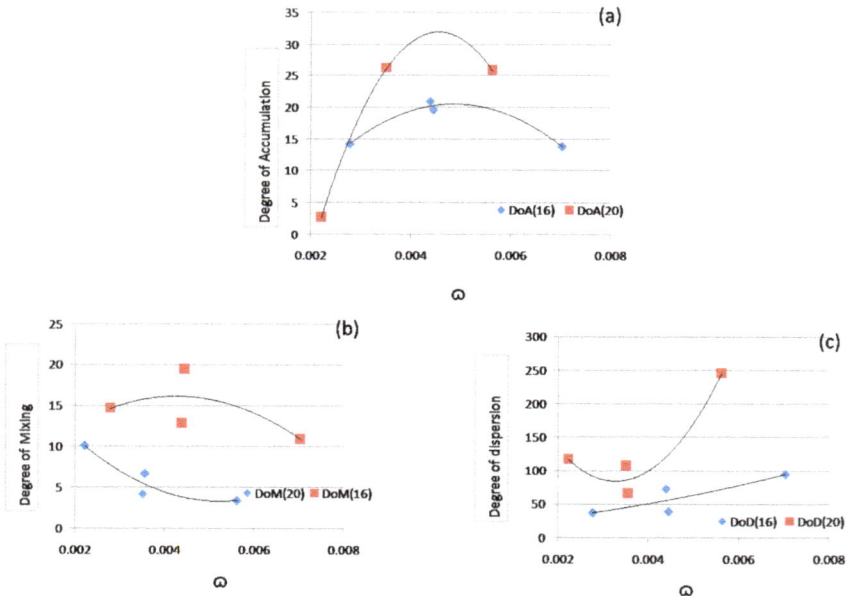

Figure 6. Effect of "ϖ" on (a) D_a, (b) D_m, and (c) D_d.

The plots given in Figure 6a–c indicate that the variation of all three indices is consistent for both diameters. In addition, an increase in unit stirring results in better dispersion, which is desirable;

further, it also causes undesirable indices to reduce. This gives valuable input to choose appropriate values of process parameters which yield maximum desirable response in the single pass FSP for SC fabrication. It is worth mentioning that an increase in shoulder diameter, increases net material movement and heat input. Under these circumstances, the nature of matrix–reinforcement interface changes, which in turn results in the change in all three indices for different shoulder diameters.

Macro- and Micro-Structure

The Macrograph and micrographs taken from the transverse section of the SC samples are shown in Figures 7–14 (i and ii). The dark grey and black portion in Figures 7–14 (ii) represents the area of reinforced zone (A_r) of the fabricated composites. The characteristic regions, namely accumulated, mixed, and dispersed regions, are clearly visible in the reinforced zone of the samples where a particular region is present. Detailed microstructure of various regions shown in the macro-scale views of Figures 7–14 (i,ii) is given in Figures 7–14 (a–d). In samples 1–4, 7 and 8, the accumulation (agglomeration) of particles is observed close to the middle in the stir zone (SZ), as shown in Figures 7–10, 13 and 14 (i and ii). The agglomeration is observed on retreating side (RS) of the SZ in Samples 5 and 6 (Figures 11 and 12). The agglomeration or accumulation occurs as a result of inadequate material flow in processed region. The main reason for the inadequate flow, which results in agglomeration, is the insufficient heat input due to undesirable FSP parameter combinations and resistance offered by reinforcement [19,21,22].

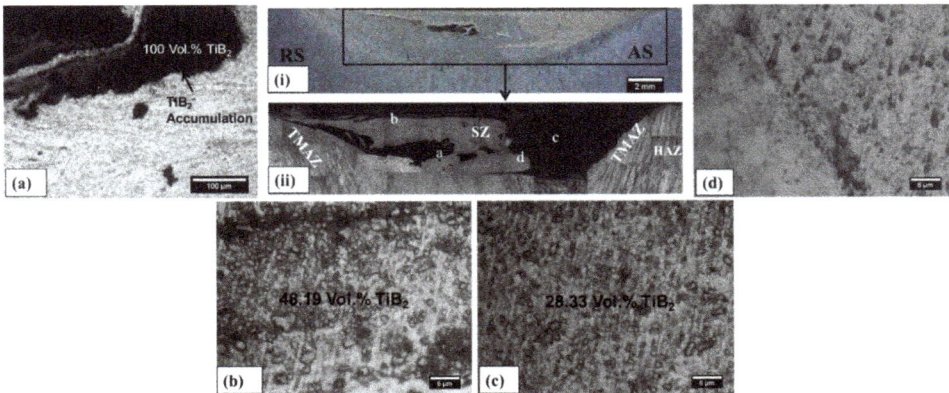

Figure 7. (**i**) Macro-image of cross-section of Sample 1. (**ii**) Microstructure of cross-section of the rectangle in macro-image showing: (**a**) agglomerated region; (**b**) mixed region; (**c**) dispersed region; and (**d**) composite and BM interface.

In this research work the processing was performed in position control mode and the plunge for kept constant. At a fixed plunge and considering the frictional characteristics for work and tool material combination to remain constant, the heat input during processing would largely depend on the ratio of tool rpm and traversing speed (T/R ratio) and on shoulder diameter. Least amount of agglomeration of particles was found in sample number 7, which may be attributed to sufficient heat input to cause efficient material flow in processed region. The density of TiB_2 particles in various regions (given in respective figures) is not same in all the regions of PZ. The distributed particles are densely packed in the mixed region; whereas, in dispersed region they are sufficiently sparsely distributed so as to have good particle-matrix bonding.

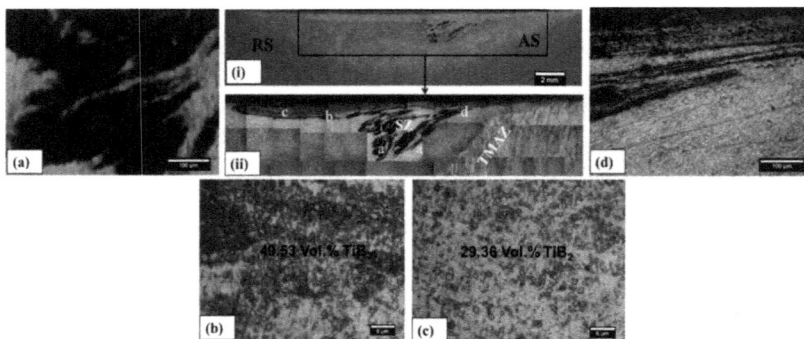

Figure 8. (**i**) Macro-image of cross-section of Sample 2. (**ii**) Microstructure of cross-section of the rectangle in macro-image showing: (**a**) agglomerated region; (**b**) mixed region; (**c**) dispersed region; and (**d**) AS interface.

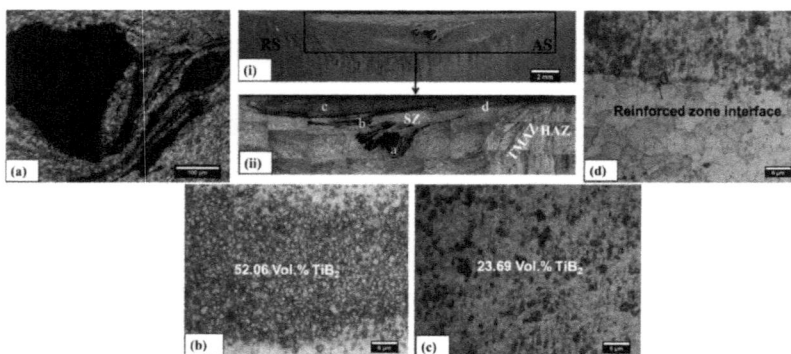

Figure 9. (**i**) Macro-image of cross-section of Sample 3. (**ii**) Microstructure of cross-section of the rectangle in macro-image showing: (**a**) agglomerated region; (**b**) partially mixed region; (**c**) dispersed region; and (**d**) reinforced zone interface.

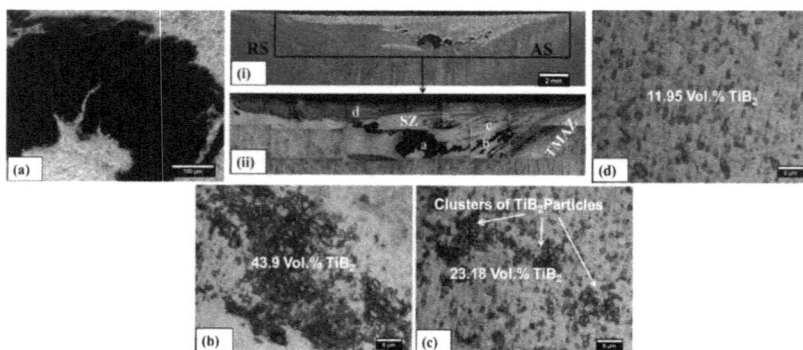

Figure 10. (**i**) Macro-image of cross-section of Sample 4. (**ii**) Microstructure of cross-section of the rectangle in macro-image showing: (**a**) agglomerated region; (**b**) mixed and Partially mixed region; and (**c**,**d**) dispersed regions.

Particle accumulation acts as stress raiser and seriously affects properties of the fabricated SCs. Mixed regions may also contain clusters of particles and may cause poor bonding with substrate material. It is necessary to achieve good bonding of reinforcement particles with the substrate effective and homogeneous particle dispersion in processed region. It is worth mentioning that the relationships between process parameters and characteristic processed regions help in avoiding the detrimental effects of these discontinuities.

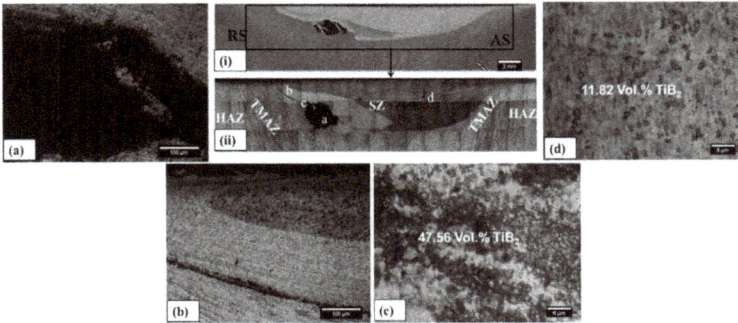

Figure 11. (i) Macro-image of cross-section of Sample 5. (ii) Microstructure of cross-section of the rectangle in macro-image showing: (**a**) agglomerated region; (**b**) RS reinforced zone interface; (**c**) mixed region; and (**d**) dispersed region.

Achieving uniform particles distribution has been a very critical issue in SC fabrication through FSP. Sustained efforts have been directed towards achieving uniform particles distribution to strengthen the material by utilizing various strategies [9,17–19,22]. In most of the available literature, it is improved by applying multiple FSP passes along the same direction of processing or reversing the direction of AS and RS [9,17–19]. Every pass reduces the accumulation of particles from spread region, and more evenly disperses the particles from mixed region. This is due to re-stirring actions of the tool, which results in homogeneous distribution of particles. Recently, Sharma et al. (2016) applied strategies such as variation in tool speeds, offset and pin diameter to achieve better distribution of SiC particles on the AA5083 by utilizing multi-pass processing technique [22]. However, every pass also simultaneously increases the processing time and net energy input, which undermines the very essence of this clean technology.

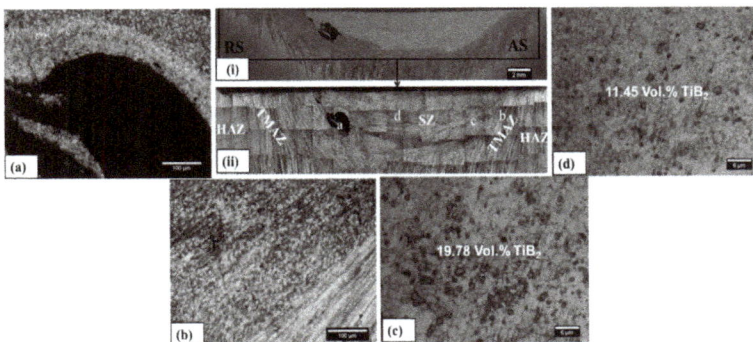

Figure 12. (i) Macro-image of cross-section of Sample 6. (ii) Microstructure of cross-section of the rectangle in macro-image showing: (**a**) agglomerated region (particles washed out during grinding and etching); (**b**) AS reinforced zone interface; and (**c**,**d**) dispersed regions.

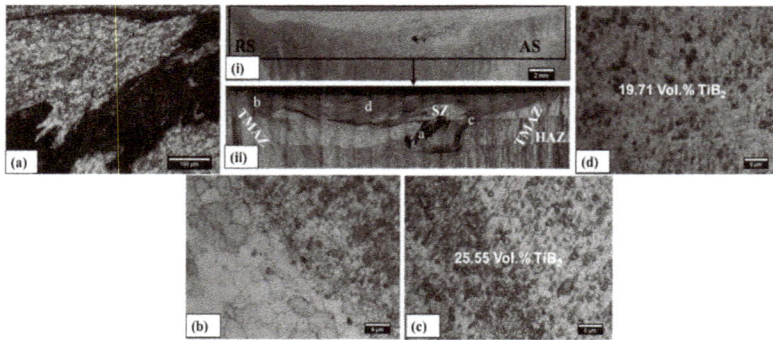

Figure 13. (**i**) Macro-image of cross-section of Sample 7. (**ii**) Microstructure of cross-section of the rectangle in macro-image showing: (**a**) agglomerated region; (**b**) RS reinforced zone interface; (**c**) mixed region; and (**d**) dispersed region.

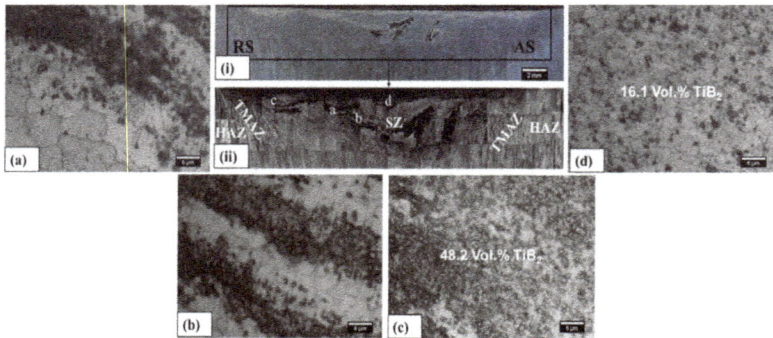

Figure 14. (**i**) Macro-image of cross-section of Sample 8. (**ii**) Microstructure of cross-section of the rectangle in macro-image showing: (**a**) RS reinforced zone interface; (**b**) mixed and unmixed region; (**c**) mixed region; and (**d**) dispersed region.

Furthermore, multiple pass FSP also requires a lot of time in setup changes. Often, flash, thinning and distortion may require substrate to be machined/corrected before being processed by subsequent passes apart from additional energy in every pass [9]. In the case of age-hardened alloys, multi-pass strategy may even prove counter-productive, as it may lead to weakening rather than strengthening. The weakening is associated with Ostwald ripening of the strengthening precipitate [19]. Above all, without understanding the mechanism of particle distribution, impromptu employment of multiple passes is not the best approach. It will be highly useful to establish an understanding on how the particles move during stirring and how the particle movement relates to process parameters. Often, the effects of important FSP parameters is contradictory; consequently, a unified/compound parameter such as that derived in the present study (i.e., unit stirring) can prove to be very useful.

The proposed derived parameter, i.e., unit stirring, is a novel single input parameter that relates effectively with all the characteristic responses (accumulation, mixing and dispersion) and also suitably controls the indirect responses, i.e., heat input and plastic deformation, thus significantly helps in predicting the overall FSP process during single pass SC fabrication. Unit stirring also guides in selecting the primary FSP parameters to achieve better distribution in a single pass.

4. Conclusions

This study provides a relation between novel derived parameter "unit stirring" and three characteristic stages of particle distribution during SC fabrication from AA7050/TiB$_2$. The study improves understanding of the relationship between unit stirring and particle distribution as well as the manner in which the particles get distributed during the course of stirring in single FSP pass. The discussed results are concluded as follows:

(1) The novel derived parameter, i.e., unit stirring, provides a useful single input parameter that relates well to the stages of stirring and helps to obtain better single pass particle distribution.

(2) Discrete stages of particle distribution (i.e., spreading with accumulation, mixing and dispersion) were identified as the mechanism of particle movement and discussed using micro- and macrographs. These stages create distinguishable characteristic regions, the size of which also relate well to unit stirring.

(3) Apart from the size of regions created in each stage, the study also proposes indices (degree of accumulation D$_a$, degree of mixing D$_m$ and degree of dispersion D$_d$) to study effectiveness of single pass FSP for SC fabrication. In addition, these indices also relate well to the unit stirring.

Author Contributions: Conceptualization, Namrata Gangil (N.G.), Sachin Maheshwari (S.M.) and Arshad Noor Siddiquee (A.N.S.); Methodology, N.G. and A.N.S.; Validation, N.G., S.M. and A.N.S.; Formal Analysis, Emad Abouel Nasr (E.A.N.), A.N.S. and Mohammed El-Meligy (M.E.); Investigation, N.G., A.N.S. and S.M.; Resources, E.A.N., A.N.S. and Abdulaziz M El-Tamimi (A.M.E.); Writing—Original Draft Preparation, N.G. and A.N.S.; Writing—Review and Editing, A.N.S. and S.M.; Project Administration, A.N.S. and S.M.; and Funding Acquisition, E.A.N., M.E. and A.M.E.

Acknowledgments: The authors extend their appreciation to the Deanship of Scientific Research at King Saud University for funding this work through research group No. (RG-1439-009).

Conflicts of Interest: The authors declare no conflict of interest.

References

1. Gangil, N.; Siddiquee, A.N.; Maheshwari, S. Aluminium based in-situ composite fabrication through friction stir processing: A review. *J. Alloy Compd.* **2017**, *715*, 91–104. [CrossRef]

2. Sudhakar, I.; Madhu, V.; Reddy, G.M.; Rao, K.S. Enhancement of wear and ballistic resistance of armour grade AA7075 aluminium alloy using friction stir processing. *Def. Technol.* **2015**, *11*, 10–17. [CrossRef]

3. Li, Y.; Qin, F.; Liu, C.; Wu, Z. A Review: Effect of friction stir welding on microstructure and mechanical properties of magnesium alloys. *Metals* **2017**, *7*, 524.

4. Goel, P.; Siddiquee, A.N.; Khan, Z.A.; Khan, N.Z.; Hussain, M.A.; Khan, Z.A.; Abidi, M.H.; Al-Ahmari, A. Investigations on the effect of tool pin profiles on mechanical and microstructural properties of friction stir butt and scarf welded aluminium alloy 6063. *Metals* **2018**, *8*, 74. [CrossRef]

5. Pao, P.S.; Gill, S.J.; Feng, C.R.; Sankaran, K.K. Corrosion-fatigue crack growth in friction stir welded Al 7050. *Scr. Mater.* **2001**, *45*, 605–612. [CrossRef]

6. Canaday, C.T.; Moore, M.A.; Tang, W.; Reynolds, A.P. Through thickness property variations in a thick plate AA7050 friction stir welded joint. *Mater. Sci. Eng. A* **2013**, *559*, 678–682. [CrossRef]

7. Jata, K.V.; Sankaran, K.K.; Ruschau, J.J. Friction-stir welding effects on microstructure and fatigue of aluminium alloy 7050-T7451. *Metall. Mater. Trans. A* **2000**, *31A*, 2181–2192. [CrossRef]

8. Huang, B.; Kaynak, Y.; Sun, Y.; Jawahir, I.S. Surface layer modification by cryogenic burnishing of Al 7050-T7451 alloy and validation with FEM-based burnishing model. *Procedia CIRP* **2015**, *31*, 1–6. [CrossRef]

9. Ju, X.; Zhang, F.; Chen, Z.; Gang, J.; Wang, M.; Wu, Y.; Zhong, S.; Wang, H. Microstructure of multi-pass friction-stir-processed Al-Zn-Mg-Cu alloys reinforced by nano-sized TiB$_2$ particles and the effect of T6 heat treatment. *Metals* **2017**, *7*, 530. [CrossRef]

10. Bahrami, M.; Dehghani, K.; Givi, M.K.V. A novel approach to develop aluminium matrix nano-composite employing friction stir welding technique. *Mater. Des.* **2014**, *53*, 217–225. [CrossRef]

11. Hashemi, R.; Hussain, G. Wear performance of Al/TiN dispersion strengthened surface composite produced through friction stir process: A comparison of tool geometries and number of passes. *Wear* **2015**, *324–325*, 45–54. [CrossRef]
12. Heydarian, A.; Dehghani, K.N.; Slamkish, T. Optimizing powder distribution in production of surface nano-composite via friction stir processing. *Metall. Mater. Trans. B* **2014**, *45B*, 821–826. [CrossRef]
13. Timoshenko, S.P. *Strength of Materials Vol. II: Advanced Theory and Problems*, 3rd ed.; CBS Publishers and Distributors: New Delhi, India, 2002; p. 698.
14. Pilkey, W.D.; Pilkey, D.F. *Peterson's Stress Concentration Factors*, 3rd ed.; John Wiley & Sons: Hoboken, NJ, USA, 2008.
15. Dolatkhah, A.; Golbabaei, P.; Givi, M.K.B.; Molaiekiya, F. Investigating effects of process parameters on microstructural and mechanical properties of Al5052/SiC metal matrix composite fabricated via friction stir processing. *Mater. Des.* **2012**, *37*, 458–464. [CrossRef]
16. Celik, S.; Cakir, R. Effect of friction stir welding parameters on the mechanical and microstructure properties of the Al-Cu butt joint. *Metals* **2016**, *6*, 133. [CrossRef]
17. Izadi, H.; Gerlich, A.P. Distribution and stability of carbon nanotubes during multi-pass friction stir processing of carbon nanotube/aluminum composites. *Carbon* **2012**, *50*, 4744–4749. [CrossRef]
18. Sharifitabar, M.; Sarani, A.; Khorshahian, S.; Afarani, M.S. Fabrication of 5052Al/Al₂O₃ nanoceramic particle reinforced composite via friction stir processing route. *Mater. Des.* **2011**, *32*, 4164–4172. [CrossRef]
19. Gangil, N.; Maheshwari, S.; Siddiquee, A.N. Multipass FSP on AA6063-T6 Al: Strategy to fabricate surface composites. *Mater. Manuf. Process.* **2018**, *33*, 805–811. [CrossRef]
20. Khan, N.Z.; Siddiquee, A.N.; Khan, Z.A.; Mukhopadhyay, A.K. Mechanical and microstructural behavior of friction stir welded similar and dissimilar sheets of AA2219 and AA7475 aluminium alloys. *J. Alloy Compd.* **2017**, *695*, 2902–2908. [CrossRef]
21. Gangil, N.; Maheshwari, S.; Siddiquee, A.N. Influence of tool pin and shoulder geometries on microstructure of friction stir processed AA6063/SiC composites. *Mech. Ind.* **2018**, in press. [CrossRef]
22. Sharma, V.; Gupta, Y.; Kumar, B.V.M.; Prakash, U. Friction stir processing strategies for uniform distribution of reinforcement in a surface composite. *Mater. Manuf. Process.* **2016**, *31*, 1384–1392. [CrossRef]

![metals logo] *metals*

MDPI

Article

Characterization of Microstructural Refinement and Hardness Profile Resulting from Friction Stir Processing of 6061-T6 Aluminum Alloy Extrusions

Nelson Netto, Murat Tiryakioğlu * and Paul D. Eason

School of Engineering, University of North Florida, Jacksonville, FL 32224, USA; netto993@hotmail.com (N.N.); paul.eason@unf.edu (P.D.E.)
* Correspondence: m.tiryakioglu@unf.edu; Tel.: +1-904-620-1390

Received: 13 June 2018; Accepted: 15 July 2018; Published: 19 July 2018

Abstract: In this study, the change in microstructure and microhardness adjacent to the tool during the friction stir processing (FSP) of 6061-T6 extrusions was investigated. Results showed that the as-received extrusions contained Fe-rich constituent particles with two distinct size distributions: coarse particles in bands and finer particles in the matrix. After FSP, Fe-containing particles exhibited single-size distribution and the coarse particles appeared to be completely eliminated through refinement. Microhardness tests showed the presence of four distinct zones and that hardness increased progressively from the dynamically recrystallized closest to the tool, outward through two distinct zones to the base material. The similarities and differences between the results of this study and others in the literature are discussed in detail.

Keywords: Vickers microhardness; Fe-containing constituents; lognormal distribution

1. Introduction

Friction stir processing (FSP) is a technique, derived from friction stir welding [1,2], where a rotating tool consisting of a pin and shoulder is plunged into the material until the shoulder contacts the outside surface of the workpiece. Subsequently, the tool is forced along the plane of the surface of the material, while the shoulder remains in contact with the workpiece. The pin forces the material to undergo intense plastic deformation, resulting in a refined, homogenized, and recrystallized microstructure [1–4]. This microstructural modification has been stated as the reason for improvement in mechanical properties, such as tensile properties and fatigue life [5–10]. Process parameters, such as rotational and transverse speed, and choice of tool geometry are critical to the material flow and to the resulting microstructural modification. Recent studies [11,12] have developed modeling methods to optimize process parameters. It should be noted that, for each material and application, there is a unique set of optimum process parameters [13]. For the current study, the authors previously developed the process parameters and the methodology in that approach which can be found in a separate publication [14]. The intent of this study was to document the phenomena of microstructural modifications, which have been largely underexplored in the literature.

Among the microstructural features that are modified during FSP are Fe-containing constituent phases [15] that form in aluminum alloys during solidification [16], and are known to reduce tensile properties [17]. It was demonstrated that the size distribution of Fe-containing constituents can be taken as the flaw size distribution in wrought aluminum alloys. DeBartolo et al. showed that the reduction of the sizes of Fe-containing constituent particles from FSP leads to smaller effective size of defects and consequently to higher tensile strength and elongation, as well as longer fatigue life [18].

Microstructural effects from FSP occur as a result of the deformation that occurs due to the stirring action of the submerged tool. The effect of the stirring action during FSP changes drastically with distance from the tool, leading to distinct zones in the microstructure from both mechanical deformation and heat dissipation zones. Woo et al. [19] characterized the microstructure of 6061-T6 alloy plates after FSP. They reported four zones: (i) the dynamically recrystallized zone (DXZ), which is the fully processed zone caused by the stirring action; (ii) the thermomechanically affected zone (TMAZ), which is generated due to the deformation and heat from the plastic deformation in DXZ; (iii) the heat affected zone (HAZ); and (iv) the base metal (BM), which is not affected by the heat. These different zones exhibit different hardness profiles. Figure 1 describes the hardness profile for each zone after different process times (after 168 hrs. and 5760 hrs. of FSP). The DXZ shows a lower hardness profile than the BM, likely due to the coarsening and/or dissolution of strengthening precipitates in the Al matrix. Minimum Vickers hardness (Hv), in Figure 1, is at the TMAZ-HAZ transition on both sides of the tool for all cases. Similar results were reported for the FSW of 6061-T6 [20]. Note in Figure 1 that the change in HV from DXZ to BM is approximately 25–30 in both cases.

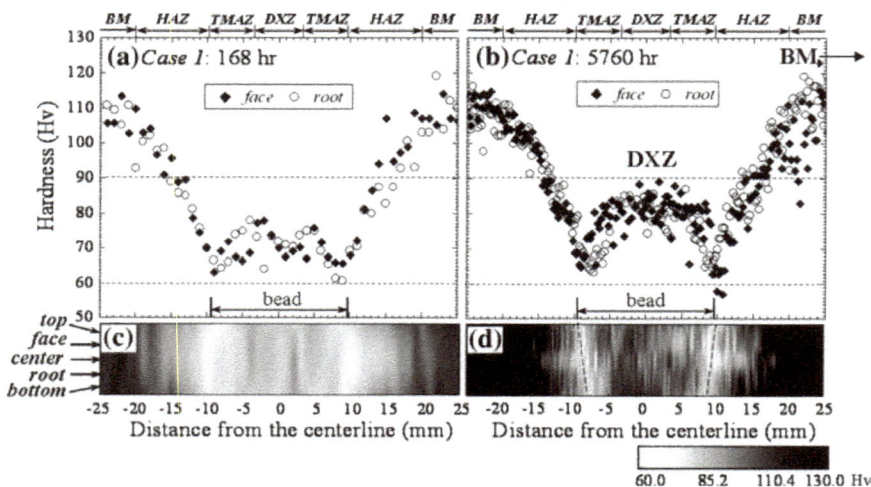

Figure 1. Hardness profiles measured along the face and root in Case 1: (**a**) 168 h and (**b**) 5760 h after friction stir processing (FSP); (**c**) and (**d**) present hardness maps, with a scale shown in the bottom right of the image [15] (US Government Work, no copyright).

2. Materials and Methods

Extruded bars of a 6061-T6 aluminum alloy with the dimensions 330×25.4 mm were used in this study. FSP was conducted on a Bridgeport vertical milling machine, with the FSP tool tilted 3° opposite to the processing direction. The tool rotation rate and transverse speed were kept constant at 700 rpm in a clockwise direction and 50 mm/min, respectively. The FSP tool was made of H13 tool steel with a shoulder diameter of 18 mm. The cylindrical pin had a diameter of 5.9 mm, a length of 5 mm, and M6-threads. After FSP, samples were sectioned by low-speed saw, mounted in epoxy, and prepared by standard metallographic polishing methods. To evaluate the hardness profile for each FSP zone in the 6061-T6 extrusion, microhardness tests were carried out on a Shimadzu HMV G21 automated Vickers microhardness tester with load of 98.07 mN and dwell time of 15 s. A Tescan Mira 3 Field Emission Scanning Electron Microscope (FE-SEM) equipped with an Oxford Instruments X-Max 50 energy dispersive spectrometer (Abingdon, Oxford, UK) was used to evaluate microstructure on unetched specimens.

3. Results and Discussion

The microstructure of the 6061 extrusion (base material) used in this study is shown in Figure 2a, which shows bands of constituent particles along the extrusion direction. The X-ray map for Fe is presented in Figure 2b, which shows that large as well as finer particles contain Fe. Moreover, these particles were also found to contain Si, which are typical of constituent particles in aluminum alloys [16]. Both large and smaller constituent particles containing Fe and Si are visible in Figure 3. The finer particles within the bands probably fractured during the extrusion process.

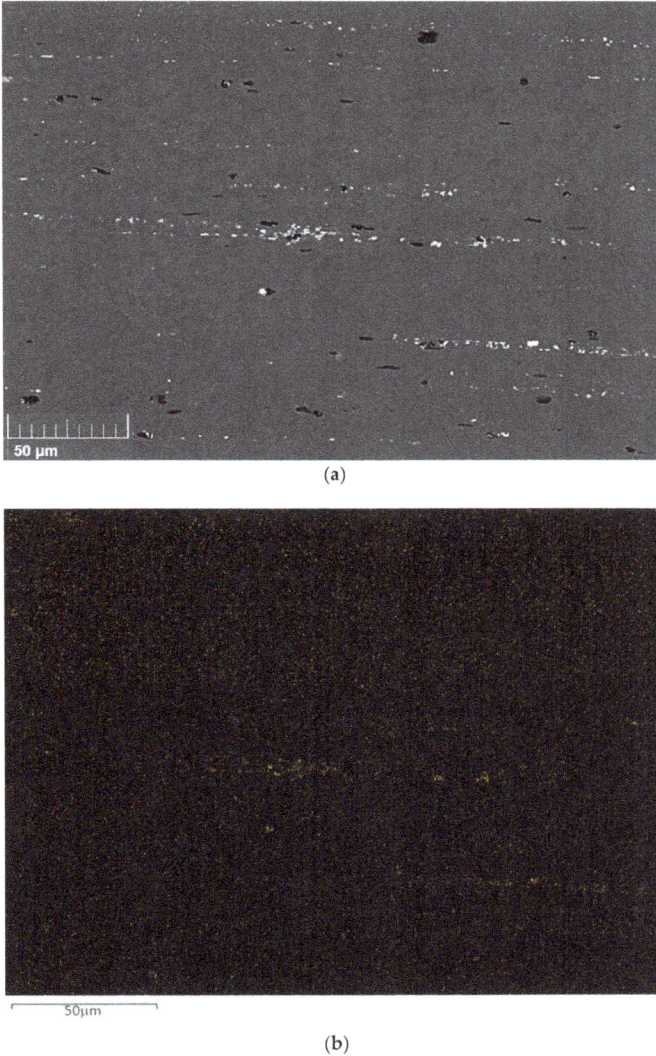

(a)

(b)

Figure 2. (a) SEM micrograph, taken in BackScattered Electron (BSE) mode, of the microstructure of the aluminum matrix and (b) X-ray map for Fe of the same region, which indicates that bright particles in the micrograph contain iron.

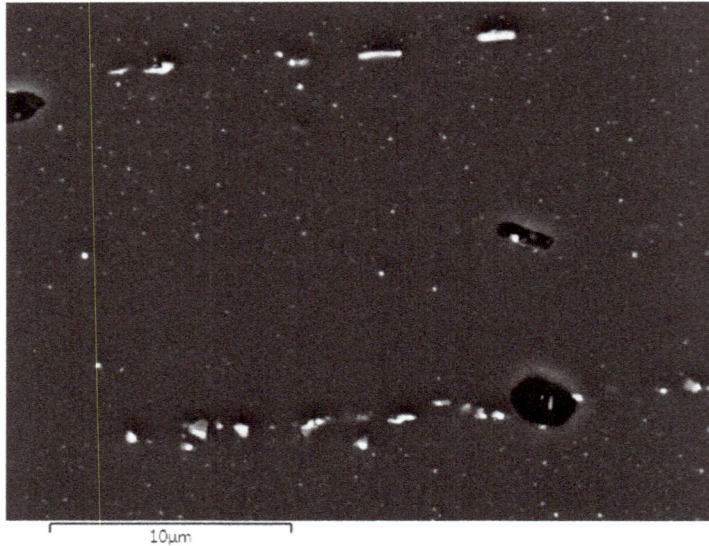

Figure 3. SEM micrograph taken in BSE Mode at higher magnification than Figure 2, showing the microstructure of 6061-T6, showing the Fe-rich constituent sizes and their distribution.

Moreover, the boundary between DXZ and TMAZ is visible at approximately 300 μm to the left from the lower tip of the tool (Figure 4). To the right of this boundary, i.e., in DXZ, the spacing of particles seems to be random. In contrast, the particles are in bands as shown in Figure 2 to the left of the boundary.

Figure 4. SEM micrograph taken in BSE mode with vertical lines added to show the boundary of the stir zone adjacent to the tool. Note: The bright particle observed above and to the left of the thread tip was determined to be a fractured piece of the tool.

3.1. Characterization of the Effect of FSP on Microstructure

To further evaluate the effect of FSP in microstructural refinement, digital image processing was conducted with ImageJ Version 1.52e free downloadable software (National Institutes of Health, Bethesda, MD, USA) to determine the sizes of the Fe-containing constituent particles. Equivalent diameter of Fe-bearing particles, d_{Fe}, was calculated as

$$d_{Fe} = \sqrt{\frac{4A}{\pi}} \tag{1}$$

Subsequently, the Fe particle size distribution for every condition was determined by hypothesizing that size follows the lognormal distribution, which is consistent with results reported in the literature [15] for β-phase (Al$_3$FeSi) platelets in aluminum alloys. The density function, f, of the three-parameter lognormal distribution is written as:

$$f(x) = \frac{1}{(x - \tau)\sigma\sqrt{2\pi}} \exp\left[\frac{-(\ln(x - \tau) - \mu)^2}{2\sigma^2}\right] \tag{2}$$

where τ is the threshold value below which probability of x is zero, μ is the location parameter, and σ is the scale parameter.

Consistent with the observations stated previously about Figure 2, it was noticed that in the BM region there were two Fe-containing particle size distributions: coarse particles in bands, and finer particles within the matrix. The probability density function (f) for the mixture of two distributions is written as [21]:

$$f = f_1 \cdot p + f_2 \cdot (1 - p) \tag{3}$$

where p is the fraction of the distribution of the lower distribution and subscript 1 and 2 refer to the lower and upper distributions, respectively. The estimated parameters of lognormal distributions are given in Table 1.

Table 1. Estimated parameters of the lognormal distributions and the fraction of each distribution in the mixture.

FSP Zone	τ (μm)	μ	σ	p	d_{Fe} (μm)
BM (fine)	0.333	−2.247	0.804	0.858	0.48
BM (coarse)	0.695	−0.291	0.772		1.70
DXZ	0.345	−2.034	1.133		0.59

After FSP, the processed microstructure in DXZ is composed mostly of small particles, as shown in Figure 5, which is a further magnification of the microstructure in Figure 4. Hence, there is strong evidence of a refined and homogeneous microstructure after FSP. The upper distribution for large particles is completely eliminated, and all Fe-containing particles are similar in size.

The probability plots of the lognormal distributions for the sizes of Fe-containing particles are shown in Figure 6. Note that the distribution for coarse particles is completely eliminated. Hence, coarse particles are broken to the size of fine particles in the extruded microstructure. This result is in agreement with the results of a previous study [15].

The mean of a three-parameter lognormal distribution is found by:

$$d_{eq} = \tau + e^{\mu + \frac{\sigma^2}{2}} \tag{4}$$

The particles sizes calculated from estimated distribution parameters are also provided in Table 1. The average particle diameter after DXZ after FSP is 0.59 μm, which is comparable to the average fine particle diameter before FSP (0.47 μm).

Figure 5. SEM micrograph taken in BSE mode from the dynamically recrystallized zone (DXZ) in Figure 4, showing refined intermetallic particles (small bright particles).

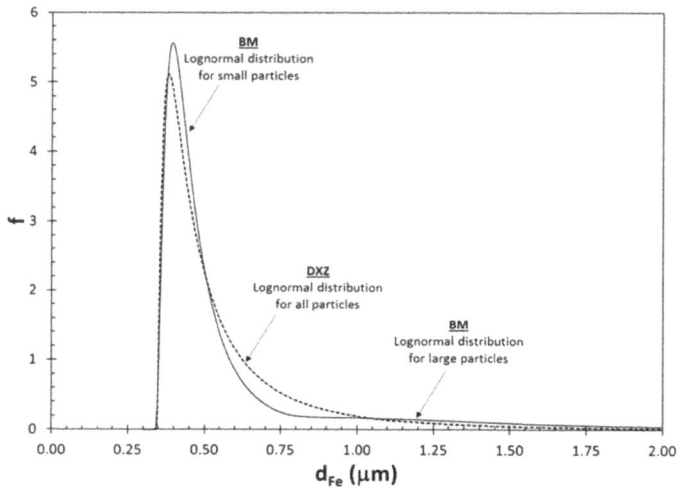

Figure 6. Fe particle size and distribution for the base material (No-FSP) and DXZ zone (FSP).

3.2. Characterization of Hardness Profile

To evaluate the hardness profile developed during FSP in the material around the tool, Vickers microhardness tests were conducted starting adjacent to the tools and moving outward at regular intervals until the base metal was sampled. One of the twenty indentations created during the microhardness tests is presented in Figure 7.

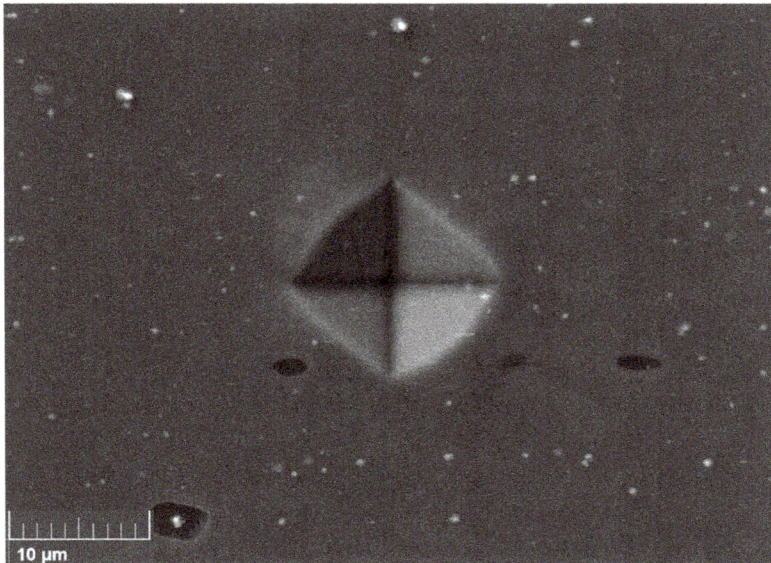

Figure 7. SEM micrograph taken in Secondary Electron (SE) mode of a Vickers microhardness test indentation in the DXZ region adjacent to the embedded tool.

The change in Vickers microhardness (Hv) with increasing distance from the FSP tool is presented in Figure 8, which shows the presence of four distinct zones, which have been interpreted to correlate with the regions suggested by Woo et al. [19]. It should be noted in the current analysis that the tool remains embedded in the sample, which is not the case in the Woo study; therefore, changes in the geometry of the regions and the extent of the DXZ must be considered. Further clarification of the variation between observed hardness in this study, and that presented by Woo et al. will be addressed later in the discussion. In the DXZ zone, the microstructure experiences greater plastic deformation than the other zones, and also results in more heat generation and lower hardness, likely the result of dissolution of strengthening precipitates in the Al matrix. Note that this zone extends from the tool surface to the boundary shown in Figure 4. In TMAZ, the reduction in material flow intensity resulted in a higher hardness profile than the DXZ. The HAZ was found to be the less affected zone by the stir action. An increase in the hardness from DXZ to the TMAZ is followed by an increase from TMAZ to the BM.

The difference between microhardness measured in the DXZ and BM in this study is 25 Hv, which is almost identical to the difference in these two zones from the Woo study, shown in Figure 1. Further comparison of the results of the two studies reveals that the widths of the zones in this study, as shown in Figure 8, are significantly smaller than those shown in Figure 1. The differences in the hardness profiles and the widths of the zones can be attributed to the differences in the process parameters in the two studies, and the presence of the remnant tool in the current investigation. Woo et al. conducted their experiments with an FSP tool that is of the same material used in the current study and with similar dimension, but using a transverse speed of 280 mm/min and rotational speed of 1250 rpm, which are both significantly higher than the levels used in the current study (50 mm/min and 700 rpm). Additionally, Woo et al. applied a compressive pressure (12.4 MPa) under the tool, which was not applied in the current study.

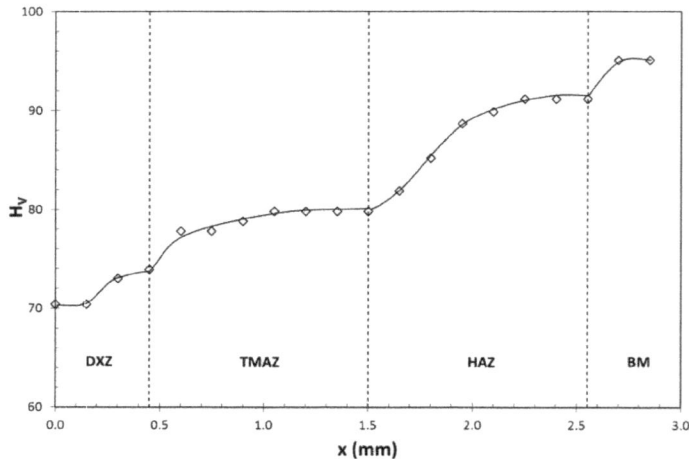

Figure 8. Microhardness profile of 6061-T6 extrusion as a function of distance from the tool edge.

4. Conclusions

- In the microstructure of 6061-T6 extrusions, there are two distinct size distributions for Fe-containing constituent particles: coarse particles in bands, and finer particles within the matrix. FSP was found to break up the large particles and refine them to the size of the initial finer particles within the matrix, which should improve mechanical properties.
- The hardness profile observed in this study was different from the one reported by Woo et al. for FSPed 6061-T6. In the microhardness profile away from the tool, four distinct zones, namely DXZ, TMAZ, HAZ, and BM, were visible. This difference was attributed to the differences in the process parameters used between the two studies, and the presence of the remnant tool in this study.
- The widths of the zones in the current study were also significantly smaller than those reported by Woo et al. These differences can also be attributed to the slower tool rotation and translation used in this study, which is presumed to result in a reduced length scale of the affected regions from the tool outward.

Data Availability: The raw/processed data required to reproduce these findings cannot be shared at this time as the data also form part of an ongoing study.

Author Contributions: The following contributions to this work were made by the following co-authors-Conceptualization, M.T., N.N. and P.D.E.; Methodology M.T., N.N. and P.D.E.; Software, M.T., N.N.; Validation, M.T., N.N. and P.D.E.; Formal Analysis, N.N., M.T.; Investigation, N.N., P.D.E.; Resources, M.T. and P.D.E.; Data Curation, M.T. and N.N.; Writing-Original Draft Preparation, M.T., N.N., P.D.E.; Writing-Review and Editing, M.T., N.N. and P.D.E.; Visualization, N.N., P.D.E.; Supervision, M.T.; Project Administration, M.T.; Funding Acquisition, none.

Funding: This research received no external funding.

Acknowledgments: The images produced for this research were all taken in the Materials Science and Engineering Research Facility (MSERF) at UNF.

Conflicts of Interest: The authors declare no conflicts of interest.

Metals **2018**, *8*, 552

References

1. Mishra, R.S.; Mahoney, M.; McFadden, S.; Mara, N.; Mukherjee, A. High strain rate superplasticity in a friction stir processed 7075 Al alloy. *Scr. Mater.* **1999**, *42*, 163–168. [CrossRef]
2. Mishra, R.S.; Mahoney, M.W. Friction stir processing: A new grain refinement technique to achieve high strain rate superplasticity in commercial alloys. *Mater. Sci. Forum* **2001**, *357*, 507–514. [CrossRef]
3. Mishra, R.S.; Ma, Z.Y. Friction stir welding and processing. *Mater. Sci. Eng. R Rep.* **2005**, *50*, 1–78. [CrossRef]
4. Jana, S.; Mishra, R.S.; Grant, G. *Friction Stir Casting Modification for Enhanced Structural Efficiency: A Volume in the Friction Stir Welding and Processing Book Series*; Butterworth-Heinemann: Oxford, UK, 2015.
5. Jana, S.; Mishra, R.S.; Baumann, J.A.; Grant, G.J. Effect of friction stir processing on microstructure and tensile properties of an investment cast Al-7Si-0.6Mg alloy. *Metall. Mater. Trans. A* **2010**, *41*, 2507–2521. [CrossRef]
6. Karthikeyan, L.; Senthilkumar, V.; Padmanabhan, K. On the role of process variables in the friction stir processing of cast aluminum A319 alloy. *Mater. Des.* **2010**, *31*, 761–771. [CrossRef]
7. Santella, M.L.; Engstrom, T.; Storjohann, D.; Pan, T.Y. Effects of friction stir processing on mechanical properties of the cast aluminum alloys A319 and A356. *Scr. Mater.* **2005**, *53*, 201–206. [CrossRef]
8. Ma, Z.Y.; Sharma, S.R.; Mishra, R.S. Effect of multiple-pass friction stir processing on microstructure and tensile properties of a cast aluminum-silicon alloy. *Scr. Mater.* **2006**, *54*, 1623–1626. [CrossRef]
9. Ma, Z.Y.; Sharma, S.R.; Mishra, R.S. Microstructural modification of as-cast Al-Si-Mg alloy by friction stir processing. *Metall. Mater. Trans. A* **2006**, *37*, 3323–3336. [CrossRef]
10. Jana, S.; Mishra, R.S.; Baumann, J.B.; Grant, G. Effect of friction stir processing on fatigue behavior of an investment cast Al-7Si-0.6 Mg alloy. *Acta Mater.* **2010**, *58*, 989–1003. [CrossRef]
11. Dewan, M.W.; Huggett, D.J.; Liao, T.W.; Wahab, M.A.; Okeil, A.M. Prediction of tensile strength of friction stir weld joints with adaptive neuro-fuzzy inference system (ANFIS) and neural network. *Mater. Des.* **2016**, *92*, 288–299. [CrossRef]
12. Chen, G.; Ma, Q.; Zhang, S.; Wu, J.; Zhang, G.; Shi, Q. Computational fluid dynamics simulation of friction stir welding: A comparative study on different frictional boundary conditions. *J. Mater. Sci. Technol.* **2018**, *34*, 128–134. [CrossRef]
13. Kumar, C.N.S.; Yadav, D.; Bauri, R.; Ram, G.D.J. Effects of ball milling and particle size on microstructure and properties 5083 Al-Ni composites fabricated by friction stir processing. *Mater. Sci. Eng. A* **2015**, *645*, 205–212. [CrossRef]
14. Netto, N.G.A. The Effect of Friction Stir Processing on the Microstructure and Tensile Behavior of Aluminum Alloys. Master's Thesis, University of North Florida, Jacksonville, FL, USA, 2018.
15. Kapoor, R.; Rao, V.S.H.; Mishra, R.S.; Baumann, J.A.; Grant, G. Probabilistic fatigue life prediction model for alloys with defects: Applied to A206. *Acta Mater.* **2011**, *59*, 3447–3462. [CrossRef]
16. Tiryakioglu, M.; Staley, J.T. Physical metallurgy and the effect of alloying additions in aluminum alloys. *Handb. Alum. Phys. Metall. Process.* **2003**, *1*, 81–209.
17. Zahedi, H.; Emamy, M.; Razaghian, A.; Mahta, M.; Campbell, J.; Tiryakioğlu, M. The Effect of Fe-rich intermetallics on the Weibull distribution of tensile properties in a cast Al-5 Pct Si-3 Pct Cu-1 Pct Fe-0.3 Pct Mg alloy. *Metall. Mater. Trans. A* **2007**, *38*, 659–670. [CrossRef]
18. DeBartolo, E.; Hillberry, B. A model of initial flaw sizes in aluminum alloys. *Int. J. Fatigue* **2001**, *23*, 79–86. [CrossRef]
19. Woo, W.; Choo, H.; Brown, D.W.; Feng, Z. Influence of the tool pin and shoulder on microstructure and natural aging kinetics in a friction-stir-processed 6061–T6 aluminum alloy. *Metall. Mater. Trans. A* **2007**, *38*, 69–76. [CrossRef]
20. Malopheyev, S.; Vysotskiy, I.; Kulitskiy, V.; Mironov, S.; Kaibyshev, R. Optimization of processing-microstructure-properties relationship in friction-stir welded 6061-T6 aluminum alloy. *Mater. Sci. Eng. A* **2016**, *662*, 136–143. [CrossRef]
21. Tiryakioğlu, M. Weibull analysis of mechanical data for castings II: Weibull mixtures and their interpretation. *Metall. Mater. Trans. A* **2015**, *46*, 270–280. [CrossRef]

MDPI

St. Alban-Anlage 66

4052 Basel

Switzerland

Tel. +41 61 683 77 34

Fax +41 61 302 89 18

www.mdpi.com

Metals Editorial Office

E-mail: metals@mdpi.com

www.mdpi.com/journal/metals

www.ingramcontent.com/pod-product-compliance
Lightning Source LLC
Chambersburg PA
CBHW051908210326
41597CB00033B/6075